T0280877

Cracking the Ad Code

Do you need to produce successful creative ideas in advertising? If so, then you need this book. For the first time, the secret of inventing new creative campaigns is unlocked, and specific, practical tools are presented to allow the quick production of creative thoughts in marketing communications.

Along with over 100 advertisement examples and numerous case studies, you also get a systematic analysis of the creation aspect of advertising, together with a taste of the real world of advertising and what makes it work. Marketing professionals in companies will learn to know what they can expect and demand from their agencies, whilst agencies will be able to explain their work to clients in an analytic language that is easy to understand.

This is essential reading for advertising professionals working for agencies and in marketing and communication departments. It is also a useful tool for students of advertising, marketing, communication, and management, from introductory level all the way up to research faculty.

Jacob Goldenberg is an Associate Professor of Marketing in the School of Business Administration at the Hebrew University of Jerusalem. His research focuses on creativity, new product development, diffusion of innovation, social networks, and complexity in market dynamics.

Amnon Levav is Co-founder and Managing Director of Systematic Inventive Thinking (SIT). For the past 14 years he has developed and facilitated innovation programs in more than 25 countries, in a wide range of organizations, among them advertising agencies such as BBDO, Leo Burnett, and McCann-Erickson.

David Mazursky is the Kmart Professor of Marketing in the School of Business Administration, the Hebrew University of Jerusalem. His research focuses on consumer behavior, creativity, product development, innovation, and interdisciplinary research relating to abstract structures in music and the arts.

Sorin Solomon is Professor of Theoretical Physics at the Hebrew University of Jerusalem and Chair of the EU Commission Expert Group on Complexity Science.

Cracking the Ad Code

Jacob Goldenberg

Hebrew University of Jerusalem, Israel

Amnon Levav

SIT – Systematic Inventive Thinking

David Mazursky

Hebrew University of Jerusalem, Israel

Sorin Solomon

Hebrew University of Jerusalem, Israel

CAMBRIDGE
UNIVERSITY PRESS

CAMBRIDGE
UNIVERSITY PRESS

University Printing House, Cambridge CB2 8BS, United Kingdom

One Liberty Plaza, 20th Floor, New York, NY 10006, USA

477 Williamstown Road, Port Melbourne, VIC 3207, Australia

314-321, 3rd Floor, Plot 3, Splendor Forum, Jasola District Centre, New Delhi - 110025, India

103 Penang Road, #05-06/07, Visioncrest Commercial, Singapore 238467

Cambridge University Press is part of the University of Cambridge.

It furthers the University's mission by disseminating knowledge in the pursuit of education, learning and research at the highest international levels of excellence.

www.cambridge.org
Information on this title: www.cambridge.org/9780521675970

First published 2009

A catalogue record for this publication is available from the British Library

Library of Congress Cataloging in Publication data
Cracking the ad code / Jacob Goldenberg ... [et al.]
    p.   cm.
ISBN 978-0-521-85905-9
1. Advertising.   I. Goldenberg, Jacob, 1962–   II. Title.
HF5821.C73   2009
659.1–dc22

                                    2008048866

ISBN   978-0-521-85905-9   Hardback
ISBN   978-0-521-67597-0   Paperback

Contents

8.   Extreme Effort   137

9.   Attribute–Value Mapping   145

Acknowledgments

Very little of the content of this book would have existed in this form without the huge collaboration of several people and organizations. Historically, the application of the Systematic Inventive Thinking method to advertising started in 1994 as a joint project of the "academics" Jacob Goldenberg and Roni Horowitz, with the "agency people" Haim Hardouf, Haim Peres, and Shuki Berg at the Symbol Peres ad agency in Tel Aviv. Shuki, the Creative Director, was our supplier of examples and industry wisdom, and the Haims believed in the project enough to support it financially in its initial stages. Roni's Ph.D. formalized, among other things, the Closed World principle which is the basis of several tools appearing in this book.

During these past 14 years, in parallel with the development of the body of research about the method, the trainers and facilitators of the SIT company (www.sitsite.com) have been applying the original tools, fleshing them out, polishing away those aspects that did not prove to be helpful, and enlarging the collection of tools and examples while working with (literally) thousands of agency people in dozens of countries. This work grounds the method in the day-to-day reality of advertising agencies and their clients, and makes all these participants in the workshops and courses true partners in its development. For this, we are truly thankful. Special thanks in this respect go to Fred Lamparter, who in 1995, as Head of Worldwide Training in Ogilvy & Mather, gave us our first chance to try out our budding tools and convinced us that, despite our shaky start, there was value in these concepts.

Since then, the main force in weaving the network of SIT users has been Martin Rabinowitz, and he is probably most responsible for its widespread acceptance. Idit Biton, Tamar Chelouche, Grant Harris, Orly Seagull, Nurit Shalev, and Yoni Stern have all been faithful teachers and helpful developers during these years, as well as the rest of the team at Systematic Inventive Thinking – SIT. Yonathan Dominitz, no longer with SIT, deserves thanks for his years as our roaming facilitator.

And last on the list, because first in their contribution to this book, many many thanks to Robyn Taragin and Omri Herzog, both from SIT. Robyn searched and researched, pled and negotiated and made sure that as many relevant examples as possible would get published; and Omri Herzog who literally transformed the varied and oftentimes messy inputs of the authors into as coherent a text as the raw materials allowed.

Introduction

Inventing tools that support advertisers in generating ideas

In the course of our research it was discovered that 89% of 200 award-winning ads fall into a few simple, well-defined design structures. This evidence may appear perplexing and difficult to account for, at least at first glance. However, this discovery and subsequent supportive research, gave us the basis to develop the set of tools we are presenting in this book. These tools support the idea generation stage in the advertising process and can be used by any advertising professional to generate his own individual, unique and creative ideas for ads.

The *creation stage* of advertising encompasses the *idea generation* process – coming up with the concept for an ad, the generation of written copy (*copywriting*), artwork of various kinds (*art direction*), and a preliminary or comprehensive version of the ad (*layout*). Many professionals view the initial phase of *idea generation* as the "heart" or "key" of the creativity process. We therefore concentrated our research on developing tools for this part of the process. We wanted to find out *how to come up with ideas for creative ads*.

Strictly speaking, there are hardly any books that focus specifically on the subject of generating ideas for advertising. There are two relevant types of book that deal with adjacent subjects. One is about improving the process of creating advertising, and touches on issues such as writing the creative brief, analyzing the market to identify insights, crafting the creative message, etc. However, this does not address the question of where the actual ideas come from. The second type of book used by ad agencies is concerned with creativity in general. Prime examples are books by Edward De Bono or books about brainstorming in its various styles. In this case, ad professionals use universal creativity techniques, and try to adapt them to their needs. "*Cracking the Ad Code*" is the only book of which we are aware that deals directly with the issue of how to *come up with new ideas* for creative ads. More specifically, this book outlines methods for actually *coming up with* the idea, methods that originate from and are designed exclusively for generating ideas *in advertising*.

The major discovery in our analysis of over 200 award-winning ads was that, amazingly, nearly 90% could be categorized into distinct patterns. Having identified these patterns, we used them to create the tools in this book: tools that can be used to develop other creative ads.

Can creativity be defined and quantified?

The notion that creativity can be defined and quantified seems a strange one to us. Is it, indeed, possible to measure a creative ad? Is it possible to define "creativity" and discover where it originates so that anyone can then generate creative ideas at will? Creativity seems to be one of the most elusive traits to define. In Popper's phrasing creativity is a divine spark that may not be dismantled and examined by use of scientific tools: "There is no such thing as a logical method of having new ideas, or a logical construction of this process. My view may be expressed by saying that every discovery contains an 'irrational element' or 'a creative intuition' ..." (Popper, 1959: 31–32).

Creativity is considered the ultimate of human qualities, central to people from all walks of life. It is one of the key measures of intelligence. Creativity is considered to be of paramount importance for all those involved in the business of advertising in general, and for copywriters and art directors in particular. Creativity is a mission of the entire advertising industry, its *raison d'être*. Successful *creativity management* is the hallmark of a vital and prosperous advertising agency. Creative thought is so valuable in advertising agencies that entire business structures are sometimes designed around the talents of one creative genius.

Oddly enough, creativity has remained a rare topic in marketing research, even though the undisputed success of many products may be attributed to consumer creativity. Likewise, it has remained relatively unexplored in the area of advertising creativity. Thinkers and researchers in the field tackle the question of where creativity comes from – is it possible to define it by rules – or is it something that can not be defined – an inspiration?

Creativity: An academic perspective

Intuitively, many believe creativity to be something that is immeasurable and unquantifiable, a view shared by many academics.

Boden (1991) suggests that "if we take seriously the dictionary-definition of creation, 'to bring into being or form out of nothing,' creativity seems to be not only beyond any scientific understanding, but even impossible. It is hardly surprising, then, that some people have 'explained' it in terms of divine inspiration, and many others in terms of some romantic intuition, or insight."

Other thinkers and researchers conclude that the secret of creativity is illuminated by the rather vague notion of *rule-transcending* rather than *rule-following*. *Rule-transcending* was defined as *total freedom*: allowing a space for idea generation through eliminating directional guidance, constraints, criticism, and thinking within bounded scopes (Csikszentmihali, 1996). Such elimination of constraints is expected to enhance the accessibility of ideas. This way, ideas can be drawn from an infinite space of ideas (Grossman, Rodgers, and Moore, 1988). Tellis echoes these views, observing that creative ideas flourish in an environment of *freedom from rules*. The simple truth about *rules*, proclaims Tellis, is that they promote *conformity* and suppress *diversity* – one of the prerequisites of creativity (Tellis, 1998: 84–85).

Creativity is described as emerging from "thin air," or even from an apparently complete "void." Sinnott argues that "it is common for a new idea to arise almost spontaneously in the mind, often seemingly out of nothing and at a time when a person may be thinking of something quite different" (Sinnott, 1959: 23). Helmholtz testifies: "[Creative ideas] often enough crept quietly into my thinking... they were simply there... But in other cases they arrived suddenly, without any effort on my part, like an inspiration... Often they were there in the morning when I awoke" (Woodworth, 1938). Poincaré (1952) describes his work on a mathematical problem in the same vein and in a casual manner: "One day, as I was crossing the street, the solution of the difficulty which had brought me to a standstill came to me all at once." Mozart likewise accounts: "When I am, as it were, completely myself, entirely alone, and of good cheer – say, traveling in a carriage, or walking after a good meal, or during the night when I cannot sleep; it is on such occasions that my ideas flow best and most abundantly" (Mozart, 1954: 34).

These thinkers, and in fact a good deal of "common wisdom" regarding creativity, suggests that creativity is impossible to quantify or to study. Yet there are scholars who disagree.

Freud observed (1990: 35) that: "There is far less freedom and arbitrariness in mental life, however, than we are inclined to assume – there may even be none at all. What we call chance in the world outside can, as is well known, be resolved into laws. So, too, what we call arbitrariness in the mind rests upon laws, which we are only now beginning dimly to suspect."

Perhaps the implication of what Freud and others are saying is that copy-writers and art directors don't have to be mentally ill or disturbed in order to be creative. Nor do they have to be outstanding geniuses. "Regular people" using common neural processes and sound creativity processes and techniques may indeed do a good job.

However, the theoretical and practical problem of deciphering the *creativity process* that leads to a *creativity product* (i.e. an *appropriate Wow!*) is as yet not resolved. Scholars adopt either the approach of rule-transcending or that of rule-dependent, with only a few entertaining the possibility that both *surprise* and *regularity* can coexist and nourish each other in the perplexing *creative process*. Boden (1995) stresses the necessary balance between *surprise* and *regularity*: "Unpredictability is often said to be the essence of creativity. But unpredictability is not enough. At the heart of creativity lie constraints: the very opposite of unpredictability. Constraints and unpredictability, familiarity, and surprise, are somehow combined in original thinking."

So, how can we invoke creativity?

If, at heart, creativity is a combination of inspiration and regulation as these experts seem to suggest – what processes can we create that will allow us to create effectively?

McIntyre (1977) aptly pointed to the interplay between *rule-following* and *rule-transcending*: "Objective rationality is to be found in knowing how and when to put rules and principles to work and when not to. Because there is no set of rules specifying necessary and sufficient conditions for large areas of practices (such as creative advertising), the skills of practical reasoning are communicated only partly by precepts but much more by case-histories and precedents."

Yet, while experience may supply many new facts which can lead our thinking, it is not a *method* (Blachowicz, 1998: 11). So, if we surmise that creativity requires this interplay between surprise and regularity, we then have the opportunity to implement a methodology to generate ideas. Trusting sheer *randomness* or pure *chance*, with utter sacrifice of rationality and better judgment, is obviously not a preferred road toward stable and continuous advertising creativity. Some methodological frameworks should be devised and implemented. The essential question remains: what method should we adopt?

Most methods for the enhancement of *idea generation* devised over recent decades have been based on the belief that in order to "ignite the creative spark," all we have to do is break away from existing mind frameworks and search diligently for the *surprising* and the *irregular* – thus reaching the aspired goal of "generating a large quantity of ideas" (Aaker *et al.*, 1992: 372). These types of methods are based on removing judgment and any constraints, thinking as widely as possible – writing down all possible ideas in the hopes that through this, people will arrive at new and different ideas that they wouldn't have previously considered. The implicit assumption behind such methods is that the greater the number of ideas, the greater the probability of achieving a qualitative set of ideas after filtering. Nobel Prize winner Jonas Pulling said: "The best way to get a good idea is to get a lot of ideas." Ideation is therefore arrived at in *quantitative* rather than *qualitative* terms. It is directed in a *random* manner.

It is our contention that these methods have, by and large, directed research and application of advertising creativity into non-fruitful and inhibiting avenues. As to the *synergetic effect* – commonly identified with unbounded randomness methods and presupposing that a group of people thinking together is superior to a "nominal group" in which individuals think alone – at least one study asserted that this plays only a minor role in creativity ideation. In a controlled experiment, ideas suggested by individuals working alone were even evaluated as superior to those raised in brainstorming sessions (Weisberg, 1992). It has been repeatedly and conclusively shown by researchers that the most prevalent method of a *brainstorming* session does not generate more ideas or greater creativity than do nominal groups (Diehl and Stoebe, 1987; 1991). All in all, groups were shown to suppress individual productivity. The quality and originality of ideas generated by groups has similarly been proven to be inferior (Paulus, Brown, and Ortega, 1999; Sutton and Hargadon, 1996).

Often, "the reason we don't see the source of our problems is that the means by which we try to solve them are the source" (Bohm, 1992: 3). The main conclusion of such mounting evidence is that an excess of ideas obscures the *ideation process*, and *randomness* and *irregularity* impede creativity. It has finally been realized that *total freedom* in idea generation is inadequate (Connolly, Routhieaux, and Schneider, 1993; Paulus *et al.*, 1993; Stroebe, Diehl, and Abakoumkin, 1992).

Creativity as a science

An inspiring attempt at creating an "exact science of creativity" was made during the 1940s by a chemical engineer named Genrich Altschuller. He postulated that there must be discernible, measurable and learnable patterns or formulas underlying successful creative ideas. By reverse engineering more than 200,000 patents and technological inventions, he succeeded in defining about 40 patterns of invention, which he labeled "standards." These patterns could be described and predicted.

The *creativity tools paradigm* takes Altschuller's ideas one step further. If we can define a pattern that unites creative ideas, then we can derive universal *tools* that characterize the evolution of successful ideas. *Creativity tools* were initially defined through reverse engineering of product innovations. The history of a product was traced through its former versions. By portraying the configuration of each product version and subsequently examining the stepwise changes between versions, we were able to observe common patterns of change which we later classified in the *creativity tools*. We found that only *five* tools cover the majority of successful new product innovations instead of 40, making application much more manageable.

The *creativity tools paradigm* strives to imbue *theoretical coherence* into the hectic, highly competitive search for the Holy Grail of advertising creativity. The *creativity tools paradigm* – with its precisely described, step-by-step methods and techniques – is not only a *theoretical construct* but also a *practical approach*, indicating effective strategies for the study and improvement of creative performance. Unlike the approach of unbounded randomness, in which the required expertise is not necessarily related to the creativity process itself, the *creativity tools paradigm* lends itself to training and has the capacity to improve creativity outcomes directly.

Under these disciplined conditions one may enjoy the benefits of a constrained, yet more fruitful and effective search for ideation. As observed by Boden (1991), "constraints – far from being opposed to creativity – make creativity possible. To throw away all constraints would be to destroy the capacity for creative thinking. Random processes alone, if they happen to produce anything interesting at all, can result only in first-time curiosities, not radical surprises."

The universality of patterns or structures and the identification of several structures across various business-related phenomena (e.g. new products, technological improvements, and advertising) suggests that there is something more general in patterns of human processing than the mere transfer of content-based knowledge. Structures represent preferred routes of processing and help the individual process and organize information by using favored processing routes that have been proven in the past to lead to productive ideas and by avoiding processing routes that do not.

The detection and utilization of structures in ads does not necessarily undermine the element of surprise that a consumer may sense when being exposed to an ad that fits a structure. Even when regularities exist, the perception of creativity is not undermined, because it still allows for the generation of ideas to which most people could not or would not have arrived (Hayes, 1978). Ads that match structures may be perceived as superior because they elicit unrecognized familiarity. They rely on structures that have been proven successful in other contexts (possibly even by the same consumers) but are nonetheless not explicitly noticeable within the new context. The inability to explicitly express or even notice the structure can also be found in judgments made by experts.

These structures involve commonalities in the way an ad is built. It does not suggest commonalities in informational content. Indeed, recent studies have revealed that ads that repeatedly use the same structure are consistently judged as highly original and favorable, and retain the benefits of *surprise*. It further appears that some design structures give rise to ideas judged as more creative than others, and evaluation of the results may be used to classify them as unique design structures. Their broad distribution and the manner in which they affect judgment resemble the generalized and fundamental rules defined by Hofstadter as *deep concepts*. Such concepts are "normally relatively hidden from the surface and cannot easily be brought into the perception of a situation; but once they are perceived they should be regarded as highly significant" (Hofstadter, 1995: 213).

So, defined structures in ads or *tools* play the role of *attractors*: paths that the self-organized mind tends to follow, assisting the individual to process and organize information by using favorable processing routes proven in the past to lead to productive ideas (see Kelso, 1997). The small number of paved routes (i.e. basic mental operations or mechanisms) avoids spending "a lot of time going down blind alleys" typical to brainstorming (Otnes, Oviatt, and Treise, 1995), and offer the much demanded "escape from freedom" which reduces anxiety (Fromm, 1971), thus maintaining – in Einstein's words – the "joy in creative expression and knowledge," and sustaining the "courage to create."

The use of *creativity tools* assures the generation of creative ideas that most people could not or would not have arrived at without them (Hayes, 1978). Experiments show that individuals trained in the *creativity tool* technique are able to generate new ideas superior to those generated by untrained individuals or people using rival techniques – as judged by experts in their fields who were blind to the existence of *tools*. Moreover, most of those tool-fostered ideas are not replicable by any other ideation technique.

Finally, *creativity tools* enable the repetition of messages, highly contributing to awareness and recall, without worrying about loss of the customers' attention or the ill effects of boredom. There is of course no guarantee that *creativity tools* will be relevant in any particular situation. However, these *tools* were assigned a high depth value precisely because they tend to crop up over and over again across many different types of situations (new products, marketing, sales promotion, and advertising creativity, etc.) and because we notice that the best insights to many problems come when *creativity tools* "fit" naturally.

Looking for a method in the creative madness

This book was born out of academic research seeking to discover if there is a discernable model behind creative ads that could be captured and classified. We found that there is.

For research purposes, and in order to create a common language, we considered an ad to be creative if it is considered to be so by advertising professionals. Practically speaking, all of the research was performed on ads that won awards in major advertising festivals. Surprisingly enough, when evaluating the ads that won these creativity awards, we were able to distill a number of distinct tools.

These tools or patterns are underlying structures or logical forms that are shared by a large number of creative ads. The basic idea is quite unintuitive. Normally, when faced with an exceptionally creative ad, our attention is drawn to the question of what makes this ad different, interesting, and exceptional. In other words, we tend to be concerned with understanding the *uniqueness* of a creative ad. For our research, on the other hand, we focused on the *commonalities* that creative ads share with other creative ads. Our research identified eight "types" of creative ads, characterized by eight patterns of creativity. This means that, according to our research, in some 70–80% of creative ads, at least one of these eight patterns can be distinctly recognized.

Expressed in a different way, our findings revealed that, beneath the surface, nearly all of the creative ads that we analyzed were, in fact, variations on the same eight themes.

The patterns that we discovered fall into two families.

- The Unification family
  - Unification
  - Activation
  - Metaphor
  - Subtraction.
- The Extreme family
  - Extreme Consequence
  - Extreme Effort
  - Absurd Alternative
  - Inversion.

Generally speaking, the first group deals with manipulating the various resources that are available for conveying a message. In this sense you can say they all deal with the medium, in a wide interpretation of the term. The second group deals more directly with the message of the ad, and more specifically – on telling the story of the message, always taking one of its key elements to the extreme.

Although we are not privy to the historical process by which an award-winning ad was really created, we have discovered that the actual result – the ad itself – follows a pattern that can be easily recognized, once you are aware of its existence. It is our assumption that the people behind creative ads apply these patterns unconsciously. In fact, we often hear from experienced agency professionals that they have certain "formulas" that they follow. This is common when creative individuals work with the same partner for a number of years during which they develop their own language and working procedures.

Once the patterns became visible, it was possible to use them for more than just classifying existing ads; the patterns can be used as guidelines for the creation of new ads. Therefore, we have transformed each of the eight patterns into a usable tool with step-by step instructions to take the advertiser from a creative brief to creative ideas that fall within the pattern. We teach these pattern-derived procedures to advertising professionals and they use them to come up with ideas.

Effectiveness of creative ads

Are all creative, award-winning ads necessarily effective? This is hotly debated in the advertising world and has been researched extensively. We deal with the task of generating creative ads, working on the assumption that, without a conclusive answer to the previous question, everything else being equal, a more creative ad is always preferable.

Integral to our knowledge of *creativity* is the notion that the *creativity product* is a construct of *novel ideas, useful* or *appropriate* to the situation. Amabile (1996: 5) determines that an outcome "will be judged creative to the extent that it is a novel *and* appropriate, useful, correct, or valuable response to the task at hand." It would be nice to think that, in some measure, "creative" awards also take into account their appropriateness for their task – how well these ads will work for the client. However, in order to arrive at more concrete answers, we have undertaken additional research that can give some weight to the effectiveness of ads that fall into the patterns that we uncovered.

After discovering the patterns, we undertook follow-up studies to determine how design structures of ads influence viewer judgment according to a number of parameters: originality, sophistication, uniqueness, and attitude towards the product. We showed participants pairs of ads for the same products – the ads were identical to each other, except that one had a tool-matching element – the element that made this ad fall into the pattern – and in the second ad, the tool-matching element was removed (Figure I.1).

(a)   (b)

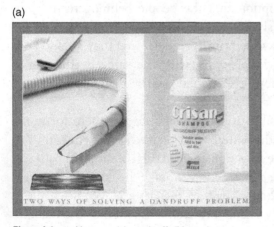
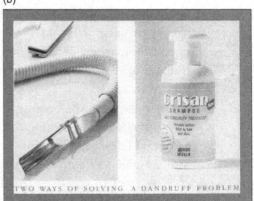

Figure I.1   Vacuum: (a) comb off; (b) comb on.

This research agenda was tested by both single ads exposure and by repeated exposure of a common design structure in five shown consecutively. The investigation shed light on the continuing efforts towards mitigating advertising wear-out, caused by repetition.

There is a noticeable pattern of results for the impact of the tool-matching features. The tool factor was significant in all cases. In all of the comparisons, the tool-matching ads were perceived as more original, unique, sophisticated, and favorable than the tool-removed ads.

The studies show that, by and large, participants do not discover the formula of the design structure, and even if they are informed about it explicitly, the ad's impact is not undermined. As a result, despite the repeated use of these structures in multiple exposures to ads with unrelated contents, creativity evaluations remain high.

An additional study was conducted in front of a computer screen. Following a brief explanation that the study was part of a marketing research project of evaluating ads, the subjects were presented with one ad at a time on the screen. After they pressed the button to indicate that the ad was understood, a new screen appeared with a short questionnaire with scales ranging from 1 to 7. The questions pertained to comprehension, attitude ("How good is the ad?"), originality, uniqueness, and sophistication. When the subject had completed the questionnaire, a new screen appeared with two open-ended questions asking their opinion of the intention of the ad, and their thoughts about how it was presented. This procedure was repeated for all 12 ads that each participant assessed.

Subjects were randomly assigned to one of three groups. In one condition participants were presented with 12 tool-matching ads – ads that included elements that matched the discovered tools. The second condition contained only tool-removed ads – ads without the element that matched the tool. In the third condition participants were presented with 12 tool-matching ads, but were also provided with a demonstration and explanation of the tool used in that particular ad and its underlying scheme. To avoid any bias due to the content of the ads themselves, the order of the ads was randomized.

As in the previous studies, the subjects' awareness of the scheme was not found to make any difference. Nor did this awareness change throughout the presentation of the 12 ads. The judgments of the originality of all 12 ads and the subjects' attitudes to them are shown in Figure I.2. It can be seen that tool-matching ads (TM) are judged as superior relative to the tool-removed (TR) ads.

There is also no indication of boredom in the subjects' judgments that could be attributable to consecutive ad judgments. This indicates that repeated

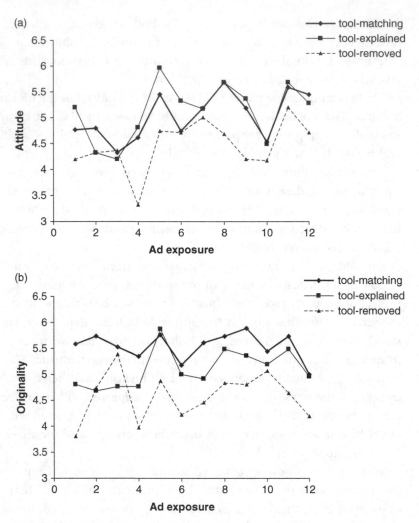

Figure I.2   Attitude and originality.

exposure to the same tool did not incur monotony. Finally, subjects were asked an open-ended question about the intention and design of the ad. The subjects did not recognize any shared element of the creative code among the tool-matching ads, even though they were presented with repeated occurrences of the same tool up to 12 consecutive times in one of the conditions. In fact, present research indicates that even when the scheme underlying the design structure was disclosed and explained; it continued to have a positive impact on creativity judgments.

The detection of underlying patterns in ads can play a role in understanding consumer reactions to ads. Research indicates that people tend to consolidate processing paths rather than proliferate them. Advertisers tend to prefer the distinguished, one-shot ad content and format, believing that it will be perceived as more creative. In contrast, recent literature, including a report on a tool-matched idea-generating algorithm (Goldenberg, Mazursky, and Solomon 1999b), indicates that structure-consistent ideas outperform random ideas and that the former are actually evaluated as more creative.

Finally, the value of studying the impact of structures in information processing can be extended to other domains. Haynes (1999) stressed the role and "generalizability" of tools:

> ... the most creative composers of music know the rules the best, and even when they break them, do so within restraints (e.g., Stravinsky decided on self-imposed limitations). Composers work with realities of sound and forms, they must relate to physical and mental structures that are enduring. This does not negate the importance of a deep understanding that may work its way into the music subconsciously. Tools help keep people's attention, e.g., a composer builds expectations and then teases them, often in small ways (e.g., themes and variations). Even big departures from expectations are often the juxtaposition of familiar elements brought together from different contexts. The 'unexpected' or hidden symmetries and tools that the listener (or viewer of art) gradually discovers are another source of pleasure associated with creativity.

A brief outline of the tools

Our research does not indicate *why* these patterns lead to more creative ads but it clearly shows the correlation to exist. However, during our 10 years of work with the patterns in the field of advertising, we have come up with some conjectures as to the reason why these structures prove to be so conducive to creativity in advertising. If we take Unification, the key is in making new and varied uses of the available resources. This means that novelty is created here when creative individuals use the medium in original ways. The specific ways are explained in the relevant chapters, but to understand the general point, let us see which resource, or aspect of the medium, is manipulated in each one of these patterns, and in what way.

- Unification – an available element of the medium or in its vicinity is used in order to deliver the message.
- Activation – the viewer is used as a "resource" in order to reveal the message.
- Metaphor – symbols or cognitive frameworks that already exist in the mind of the viewer are exploited in order to deliver the message.
- Subtraction – elements of the medium considered to be indispensable are excluded.

The secret of these patterns is, therefore, that they present the viewer with something arresting and surprising through the novel and unexpected manipulation of available resources.

The Extreme family is a descendant of the natural selling impulse, so to speak. The basic, most intuitive thing to do when you want to convince someone to buy your product, is to tell them what wonderful things this product will do for them. This is what we call "Extreme Promise." Note that this is *not* one of the patterns on the list above. The reason is that this type of advertising has already been abused to the point that no sophisticated consumer will allow it to pass through his or her defenses. Extreme Promise tells the potential purchaser: "Buy my drink because it will make you very very happy." This does not work anymore, and cannot be classed as creative advertising. The Extreme family is, therefore, a collection of patterns that tell you, in a variety of ways, why we are *not* trying to sell to you. That is why the results are considered creative and why they work well in penetrating the viewer's defenses. Again, the patterns are demonstrated and explained in detail in the corresponding chapters, but the gist of their non-selling character can be understood in the following way.

- Extreme Consequence – presents a very extreme and sometimes negative situation that happens as a consequence of using the product.
- Extreme Effort – presents the exaggerated efforts a company will go to in order to please the customer, or the absurd lengths a consumer would go to get hold of the product.
- Absurd Alternative – presents a possible, albeit highly outlandish and impractical, alternative to the product being offered.
- Inversion – suggests just how horrible the world would be without the advertised product.

When executed properly they all imply, of course, that the prospective customer should buy the advertised product, although they never articulate this explicitly.

In addition to the eight tools covered in this book, we will make constant reference to two principles, which are complimentary to the practical application of the tools.

Fusion

Within the context of advertising, Fusion is the process by which you meld into one both the symbol for something – its story – and the product or brand you are marketing and/or the advertising message. So for example, if you are marketing a fast-food chain and your message is that it provides healthy food – you could choose to express this by showing a healthy kid eating the food. That would not be fusion, as the child and the product are distinct – they are two separate images. However, if you turn the fast-food branch into a tomato – that would be fusion.

Our research has concluded that the greater the fusion that is undertaken between the message, the product and the symbol – or story brought in to the ad, the more the ad will be seen as creative.

Closed World

When choosing symbols or ideas on which to build your advertising message, you can choose to include elements that come from what we call "the closed world of a product" – which means that they have something to do with the product or its message. Examples for this would be – if we are advertising a fast-food chain, elements from the closed world of the product would be hamburgers, fries, drinks, a cow, tables, smiling waiters, etc. Your other options would be to include exterior elements. These can include an elephant, a green meadow, a boat, etc. Our research has concluded that ads that use elements from the closed world of a product are considered more creative.

BIBLIOGRAPHY

Aaker, D. A., Batra, R., and Myers, J. G. (1992). *Advertising Management*. London: Prentice-Hall International.

Altschuller, G. S. (1985). *Creativity as an Exact Science*. New York: Gordon and Breach.

Altschuller, G. S. (1986). *To Find an Idea: Introduction to the Theory of Solving Problems of Inventions*. Novosibirsk, USSR: Nauka.

Amabile, T. M. (1983). *The Social Psychology of Creativity*. New York: Springer-Verlag.

Amabile, T. M. (1996). "A Model of Creativity and Innovation in Organizations." In Amabile, T. M. ed. *Creativity in Context*. Boulder, CO: Westview Press.

Blachowicz, J. (1998). *Of Two Minds: The Nature of Inquiry*. Albany, New York: State University of New York Press.

Boden, M. A. (1991). *The Creative Mind: Myths and Mechanisms*. London: Abacus.

Boden, M. A. (1995). "Creativity and Unpredictability." *SEHR*, **4** (2).

Bohm, D. (1992). *Thought as a System*. London: Routledge.

Connolly, T., Routhieaux, R. L., and Schneider, S. K. (1993). "On the Effectiveness of Groups Brainstorming: Test of One Underlying Cognitive Mechanism." *Small Group Research*, **24**: 490–503.

Csikszentmihali, M. (1996). *Creativity*. New York: Harper Collins Publishers.

Diehl, I. M. and Stoebe, W. (1987). "Productivity Loss in Brainstorming Groups: Toward the Solution of the Riddle," *Journal of Personality and Social Psychology*, **53**: 497–509.

Diehl, I. M. and Stoebe, W. (1991). "Productivity Loss in Idea-generation Groups: Tracking Down the Blocking Effect." *Journal of Personality and Social Psychology*, **61**: 392–403.

Einstein, A. "Letter to Jacques Hadamard." In: Ghiselin, B. (ed.), *The Creative Process: A Symposium*. Berkeley and Los Angeles, CA: University of California Press, pp. 32–33.

Finke, R. A., Ward, T. B., and Smith, S. M. (1992). *Creative Cognition: Theory, Research, and Applications*. Cambridge, MA: The MIT Press.

Freud, S. (1990). *Art and Literature*. London: Penguin Books.

Fromm, E. (1971). *Escape from Freedom*. New York: Avon Books.

Goldenberg, J. and Mazursky, D. (2002). *Creativity in Product Innovation*. Cambridge: Cambridge University Press.

Goldenberg, J., Mazursky, D., and Solomon, S. (1999a). "Creativity Tools: Towards Identifying the Fundamental Schemes of Quality Ads." *Marketing Science*, **18**: 333–351.

Goldenberg, J., Mazursky, D., and Solomon, S. (1999b), "Creative Sparks." *Science*, **285**: 1495–1496.

Grossman, S. R., Rodgers, B. E., and Moore, B. R. (1988). *Innovation, Inc.: Unlocking Creativity in the Workplace*. New York: Wordware Publishing.

Gruber, H. E. and Wallace, D. G. (1999). "The Case Study Method and Evolving Systems Approach for Understanding Unique Creative People at Work." In: Sternberg, R. J. (ed.), *Handbook of Creativity*. Cambridge: Cambridge University Press, pp. 93–115.

Guilford, J. P. (1967). "Creativity." *American Psychologist*, **5**: 444–454.

Hayes, J. R. (1978). *Cognitive Psychology: Thinking and Creating*. New York: Dorsey Press.

Haynes, D. (1999). "Tools are Central to Creativity." *Science*, Published Debate Responses, September 13.

Hofstadter, D. R. (1995). *Fluid Concepts & Creative Analogies: Computer Models of the Fundamental Mechanisms of Thought*. New York: Basic Books, 1995.

Kasof, J. (1995). "Explaining Creativity: The Attributional Perspective." *Journal of Creativity Research*, **8** (4): 311–366.

Kelso, J. A. S. (1997). *Dynamic Patterns: The Self-Organization of Brain and Behavior*. Cambridge: MIT Press.

Kover, A. J., Goldberg, S. M., and James, W. L. (1995). "Creativity vs. Effectiveness? An Integrating Classification for Advertising." *Journal of Advertising Research*, **35** (6): 29–40.

Lubart, T. I. (1994). "Creativity." In Sternberg, R. J. (ed.), *The Nature of Creativity*. New York: Cambridge University Press, pp. 289–332.

Lumsden, C. J. (1999). "Evolving Creative Minds: Stories and Mechanisms." In: Sternberg, R. J. (ed.), *Handbook of Creativity*. Cambridge: Cambridge University Press, pp. 153–168.

Maimon, O. and Horowitz, R. (1999). "Sufficient Condition for Inventive Ideas in Engineering." *IEEE Transactions, Man and Cybernetics*, **29** (3): 349–361.

Martindale, C. (1999). "Biological Basis of Creativity." In: Sternberg, R. J. (ed.), *Handbook of Creativity*. Cambridge: Cambridge University Press, pp. 137–152.

McIntyre, A. (1977). "Epistemological Crises, Dramatic Narrative and the Philosophy of Science." *Monist*, **60**: 453–472.

Mozart, W. A. (1954). "A Letter." In: Ghiselin, B. (ed.), *The Creative Process: A Symposium.* Berkeley and Los Angeles: University of California Press, pp. 34–35.

Mumford, M. D. and Gustafson, S. B. (1988). "Creativity Syndrome: Integration, Application and Innovation." *Psychological Bulletin*, **103**: 27–43.

Ochse, R. (1990). *Before the Gates of Excellence: The Determinants of Creative Genius.* Cambridge: Cambridge University Press.

Oldham, G. R. and Cummings, A. (1996). "Employee Creativity: Personal and Contextual Factors." *Academy of Management Journal*, **39** (3): 607–635.

Otnes, C., Oviatt, A. A., and Treise, D. M. (1995). "Views on Advertising Curricula from Experienced 'Creatives'." *Journalism Education*, **49**: 21–30.

Paulus, B. P., *et al.* (1993). "Perception of Performance in Group Brainstorming: The Illusion of Group Productivity." *Personality and Social Psychology Bulletin*, **19**: 78–89.

Paulus, P., Brown, V., and Ortega, A. (1999). "Group Creativity." In Purser, R. E. and Montuori, A. (eds.), *Social Creativity*, Vol. 2. Cresskill, NJ: Hampton Press.

Poincaré, J. H. (1952). *Science and Method.* New York: Dover.

Popper, K. (1959). *The Logic of Scientific Discovery.* New York: Basic Books.

Ray, M. and Myers, R. (1986). *Creativity in Business.* Garden City, New York: Doubleday.

Sinnott, E. W. (1959). "The Creativeness of Life." In: Anderson, H. H. (ed.), *Creativity and Its Cultivation.* New York: Harper & Row, pp. 12–29.

Sternberg, R. J. (1988). *The Nature of Creativity: Contemporary Psychological Perspectives.* Cambridge: Cambridge University Press.

Sternberg, R. J. and Lubart, T. (1995). "Investing in Creativity." *American Psychologist*, **51**: 677–688.

Stroebe, W., Diehl, M., and Abakoumkin, G. (1992). "The Illusion of Group Productivity." *Personality and Social Psychology Bulletin* **18** (5): 643–650.

Sutton, R. I. and Hargadon, A. (1996). "Brainstorming Groups in Context: Effectiveness in a Product Design Firm." *Administrative Science Quarterly*, **41**: 685–718.

Tellis, G. J. (1998). *Advertising and Sales Promotion Strategy.* Massachusetts: Addison-Wesley.

Unsworth, K. (2001). "Unpacking Creativity." *Academy of Management Review*, **26** (2): 289–297.

Weisberg, R. W. (1992). *Creativity Beyond The Myth of Genius.* New York: W. H. Freeman.

Woodworth, R. S. (1938). *Experimental Psychology.* New York: Holt.

1 Unification

Where do ideas come from?

We ran a seminar for European advertising professionals. During the seminar, we asked participants to come up with an idea for advertising beer inside a bar. We divided participants into pairs and gave each pair 6 minutes to complete the task.

When the time was up, success was limited. A significant number of the pairs hadn't come up with any idea at all. Among the ideas that were raised, none were especially creative.

Then we tried a different tactic. We began the second part of the exercise by asking participants to list the elements of a pub. Within minutes, we had compiled a detailed list of bar elements: bar, sink, waiters and bartender, coasters, clothing that people in the bar wear, mirror, toilets, floor, dartboard, and more. With the list complete, we arbitrarily matched each pair with one element from the list. Then, we instructed each pair to come up with an idea for advertising beer, using only that one element that they were given to deliver the message. We gave each pair 5 minutes to complete this task.

That's when the floodgates opened. The increase in the quantity of the output and its quality was remarkable – and impressive. When the time was up, every single pair had not only one idea but several. Some of these ideas were even especially interesting and novel. At this stage, group members devised the idea for an advertisement in the bathroom: they would place an elegant, opulent urinal next to the regular urinals. Painting it gold and decorating it with stunning lighting fixtures, they would place a sign reading: "This exclusive urinal is reserved for Stella Artois drinkers. Stella Artois, the beer known as 'reassuringly expensive'." Another idea that was raised was using the beer glasses themselves as an advertising medium. When the drinker would be almost at the end of his beer, he would discover text on the bottom of the glass that could only be read when the glass was almost empty: "Isn't it time to order another one?"

So how can we explain the participants' relative failure in completing the first task and their success in the second? We generally relate expressions like "free thought" or "associative thinking" to creative work. In this case, however, it was specifically the exercise that demanded "controlled" thinking on the medium that delivered much greater output. The example above forces us to re-evaluate the function of constraints, limitations and schemes on the creative thinking process. It seems that far from being a drawback or a restriction, constraints can be used to true advantage and as a valuable resource in idea generation.

Prisons that set you free

In order to understand how constraints work in idea-generative thinking, let us look at where else they are found and how they operate.

Constraints can be found in all forms of creative work – and are prevalent in the work and in the world of advertising. For example, a client insists upon certain constraints like: message, advertising strategy, choice of target audience and/or format. Other constraints can be dictated by the character of the client. Some would require a conservative campaign, whereas others would prefer a more creative tack; one company would do better with a campaign that's rolled out in stages while another requires stand-alone advertisement. Budget and manpower pose natural constraints, as do restrictions of each individual medium, which we often look upon as immutable forces. Every advertising campaign necessitates two central constraints: the struggle for an audience's attention and the need to relay a clear and focused message.

These constraints are an integral part of the work of advertising: advertisers must consider these issues in order to be effective in their final product.

However, how do these constraints affect creative thinking? As discussed in the Introduction, by remaining within the closed world of an advertisement, it is possible to focus thinking and arrive at better solutions. Explained another way, by remaining within the limitations or constraints posed by each individual project, advertisers can use these constraints as guidelines for thinking and achieve more creative results. Constraints actually assist in focusing the activity of thinking and thus are able to assist in getting better results.

The constraint of the medium

In this chapter we are going to look at one particular constraint – the constraint of the medium – and discover how taking advantage of this constraint, rather than feeling trapped by it, has been proven to deliver excellent results. Media can be described as the tool selected to carry the advertising message: television, flyers, billboard, radio, etc.

In order to begin this dialogue, however, let us look at the specific constraints of one medium – billboard advertising.

In billboard advertising, or more specifically – in advertising that does not utilize space allotted by electronic or written media, but exists "out in the world," additional constraints necessarily come into play. Here are a few that stand out the most:

(1) time of exposure, usually short;
(2) changing format;
(3) varying distances between the advertisement and the audience;
(4) "one-shot" monologue – an inability to offer ongoing messages, or to offer effective continuity of advertising messages.

These constraints make billboard advertising an especially interesting challenge. The advertiser must be immediately effective as he has only a moment to create an impression. To accomplish this, the advertiser must relate to a new set of constraints.

(1) The advertisement must be simple, with only one dominant element.
(2) The design must be approachable and clean; it can include a few lines of text in large type, with frequent use of contrasting colors.

Paradox: Mixing constraints with daring

Although it is easy to see how meeting the demand of these criteria is essential to creating a good advertisement, it is not enough. In order to turn a good advertisement into a truly successful campaign, the advertisement must also be daring.

This truth poses a bit of a problem for an advertiser. He must work within the rules, but then he must also dare. Where is the line?

Advertisers must dare, within – and by understanding – the constraints that are set. Understanding the rules, respecting what they provide, he learns how and when to do something differently – he learns how to dare effectively.

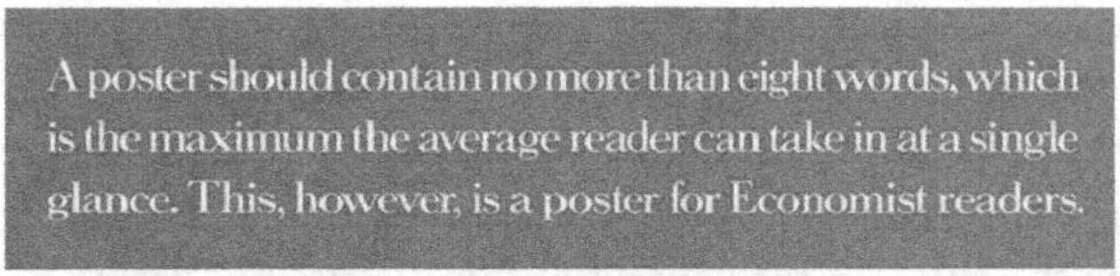

A poster should contain no more than eight words, which is the maximum the average reader can take in at a single glance. This, however, is a poster for Economist readers.

Figure 1.1   Eight word limit.

An example of this is presented in Figure 1.1, which is the billboard advertisement for *The Economist* that won the Cannes OneShot Advertisement Contest, 2001. This billboard advertisement featured a long text explaining that effective billboard advertisement shouldn't exceed eight words, which is the amount an average reader can absorb when looking briefly at something. This advertisement, of course, broke its own rule; and this was its daring – it deviated from the rule it had laid out in order to stress the message that *The Economist* readers were certainly no "average reader." However, the daring worked, because they implemented this strategy in the context of the constraints. The magazine has been using a unified graphic line in their advertising for many years: white text on a red background with a unified font. As the design format had remained unchanged for so long, *The Economist* advertisements had become easily recognizable even before you read the brand name. They were able to use their "moment" of the audience's attention in a daring way.

However, it is this desire to dare in an advertisement that makes constraints problematic. Otherwise, we could simply follow rules. The above example is an excellent demonstration of this: no one would dispute the effectivity of the constraint of using very short text in billboard advertising. However, only because of his "daring", and only because of a very particular situation, the advertiser chose not to follow the rule. Advertisers clearly have to obey constraints in order to turn out a professional advertisement. Yet, in order for an advertisement to stand out as memorable and truly effective, the rules must be broken. *This is the quintessential advertising paradox: If you want to create a successful advertisement, you must follow the recipe of certain constraints. However, if you do follow all the rules, your advertisement will be considered similar – perhaps too similar – to all other advertisements.* In a similar vein: it is easy to deviate from the constraints of creating a good advertisement. Anyone can do it. However, will deviating from the rules mean that one has necessarily implemented the quality of daring? What actually defines daring?

It is possible to walk the line. There's a simple rule: constraints are balanced effectively by daring when the constraint itself (in our case, the medium) or its components are used in a completely new way. In the context of the medium, the guideline can be expressed as follows: use the medium itself – and its components – in a completely new way in order to achieve daring results. This also gives us a clear answer regarding the definition of "daring." Daring isn't just breaking the rules. In the context of this discussion, daring is defined by using the medium or its components in a completely new and surprising way. When approaching an advertisement, discover a new way to use the medium itself or its components to tell the message you are trying to convey. This describes the tool of Unification.

Definition of advertising media

Advertising media is any platform that is used to deliver an advertising message. This can be television, radio, billboards, a DVD, the county fair, the side of a bus, placards at a football stadium, T-shirts – or any medium an advertiser utilizes to deliver his advertising message.

The elements of a medium are the components that make up that medium – a newspaper is made up of paper, ink, staples, newsprint, photographs, folds; a bus has passengers, doors, windows, color, texture, and more.

Using the medium

Old and new media

To implement the tool of Unification, the medium chosen as a platform for the advertisement is used creatively in order to deliver the message. Any media that you choose or invent will work. "Creatively using" the medium means taking advantage of the elements that already exist in that media when seeking ways to deliver a message. (If the medium is a bus, you would consider using its doors; if the medium is a skywriter you'd consider giving new meaning, perhaps, to the color of the smoke that comes out of the plane.) The tool of Unification is not based on inventing a new advertising media, although creating new media sometimes proves useful, as will be seen in the example below. By using the tool of Unification, advertisers give new life to a recognized advertising medium by using elements of that medium unconventionally in order to deliver a message. Advertisers can discover new media to

Figure 1.2   If it hurt to pick this up, use it to call us.

use as advertising platforms by paying careful attention to the regular way people operate. The idea behind any use of elements of media is always to serve the desired message.

Here is an example. A few years ago, a large number of quarters were scattered on the sidewalks of Chicago in the middle of the night. The next day, passers-by hurrying to work noticed a promising glitter on the asphalt and picked up the coin. But they were in for a surprise. The quarter was inscribed with the following message: "If it hurt to pick this up, use it to call us." The text was complemented by the telephone number of a chiropractic center. The coin itself, like all quarters, was usable in phone booths (Figure 1.2).

Making new use of existing resources

Unification is a tool that assists the advertiser in thinking systematically so that he may locate and use media in a new and unconventional way. Another important advantage of the tool is the conservation of resources used in designing a new medium or in using an old one. The creative use of the medium is not based on establishing a short-term grandiose platform that will amaze the clients and be forgotten. Through analyzing advertisements that have won creativity awards it was discovered that sometimes, making new use of an existing resource can actually be considered brilliant. More importantly, making new use of an old resource creates the effect that the advertisement will be remembered for a long time.

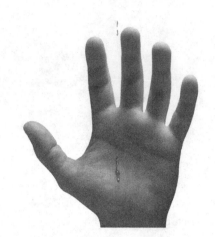

Figure 1.3   Using the staple.

Even incredible advertisements aren't usually based on the creation of new or unfamiliar media; they use recognized and familiar media in a new and creative way. The medium is generally viewed as a given resource. It is defined and taken for granted. Most of an advertiser's efforts are devoted towards improving or sharpening the message that appears in the medium. But the tool of Unification allows us to devote creative thinking to the medium itself and to try and improve it or stress specific aspects that serve the objectives of the advertisement.

Consider now Figure 1.3, in which a band-aid is advertised on a two-page spread.

Although advertising an everyday, recognizable product, the advertisement grabs the viewer's attention; viewers will most probably spend more time looking at it than they do at an average advertisement. It is reasonable to assume that this advertisement will be remembered for a long time. The advertisers created this effect by employing the services of a simple component that already exists in the medium – the staple that holds the pages of the newspaper together. The reader doesn't usually even notice this component. We run into the staple so many times that we forget it exists. The staple is a headache – or a serious drawback to most newspaper graphic artists; it is a component that is built into the medium and harms its graphic perfection. Generally, their attitude towards the staple would be to ignore it, since there is nothing they can do about it. It is a "constraint" dictated by the medium itself. However, taking this modest and ignored staple and making it the physical and conceptual center of the advertisement is surprising and creative. The advertisement forces a meeting between the viewer and the staple – and because the staple is positioned as a new concept, we discern it and give it special attention; in this way, we also pay more attention to the advertising message that is behind it.

Resources of the medium as opportunity

The above was a truly excellent example of how manipulating the medium turns its elements into advantages. This advertisement is considered creative because the advertiser did not look at the staple – or the element of the medium used – as a constraint or a drawback, but as an effective opportunity to advance the message. The new meeting with the medium makes us look again at what we recognize – what we know. It forces us to refresh the way we look at it. It forces us to see the staple. Creative advertisements that are based on this tool use the medium itself to make the viewer see the medium in a new light. The component of the medium does not fill its usual "transparent" task, but instead, acts surprisingly like an element of a particular message, which causes the viewer to give it his complete attention. This effect will reoccur every time the viewer comes into contact with the advertisement and thus, the ad will remain in the mind of the viewer for a very long time.

In Figure 1.4, the advertiser made use of a component of the medium that is not usually used. The doors of the bus meet for a kiss. The bus drives up and down the streets all day – and each time the bus stops, the kiss occurs again and again. And each time the couple meets for a kiss, they promote the message of fresh breath – delivered by the mint candies in the advertisement.

Here is another – completely different – example of use of this same component in the medium "bus" (Figure 1.5).

(a)

(b)

Figure 1.4   Kiss on the bus.

Figure 1.5   Shark's teeth.

Drawing on resources that surround the medium

The two examples above draw on resources that would exist in the medium in any case. The advertisements take this component and use it in a way that serves the message. Unification is not only limited to this. This tool can be applied to calling on resources in the environment of the medium as well. As an example, the series of advertisements for the local Swedish paper uses components located in the environment of the medium (billboard) in order to emphasize the paper's involvement in the life of the city. The message promotes the paper as knowing everything that is happening – including burnt out street lights right near one billboard; a tree recently trimmed right near another billboard; or the telephone number of the public phone on the corner – and even reporting it to its readers (Figure 1.6).

(a)

(b)

(c)

Figure 1.6   (a) That street light needs changing; (b) just trimmed; (c) call here.

Two approaches to using Unification

The tool of Unification confers a new task on an existing resource – either within the medium or in the environment of the medium – in order to demonstrate the message. In other words, the tool uses an existing component of the medium or of its environment in a way that demonstrates the problem or the promise set out in the message.

There are two strategies for using this tool, each engendering a completely different cognitive approach. In the first approach, the advertiser chooses a given medium – television, billboard, newspaper, quarters in front of the supermarket – and explores which components can assist in delivering the product's message in a more interesting, deeper or more daring way. The first approach of Unification says – you have the medium; let us use it creatively to deliver the message. However, the second approach to Unification comes from the opposite thinking process. The advertiser starts with a message. He then seeks the appropriate medium based on the message he has. Both approaches to the tool are effective and useful – and the advertiser should choose the method best suited to each particular campaign at hand. In order to use each approach, we will explore the thinking process that supports it.

Approach 1: Exploring the medium to serve the message: Getting more out of an existing medium

Segmentation and the medium

Segmentation is one of the criteria that can be used to guide which medium is selected and how a medium is used. For example, when planning a television advertising campaign, an advertiser selects airtime based on the market segments that view a particular program; advertisers seek to speak directly to the audience that is most appropriate. In this context, segmentation is a deciding factor in how the medium is used. Similarly, a particular medium can be selected in order to approach different market segments – thus, segmentation affects choice of media.

Speaking directly to your particular audience

CareerBuilder, an American human resource placement company, chose to advertise on buses. On the surface, it would seem that they chose a traditional

Figure 1.7   Don't jump.

medium. However, the company's advertising team chose to place the advertisement on the roof of the bus. Clearly, the advertisement's exposure would be significantly reduced, but they aimed for a very focused and defined market segment: bored workers who were looking out the window when taking a break from their unhappy jobs (Figure 1.7). So, although the viewing percentage was lower, the people who did see it were exactly the ones for whom it was intended and thus the strategy was effective.

In the following example, a billboard was chosen as the medium of choice to advertise a basketball club. The billboard, naturally, hung at quite a height off the ground. A basket was affixed to the sign and flyers providing information about the new club were placed inside. However, the only passers-by who could pull a flyer out of the basket were those who were especially tall (Figure 1.8). In this case, the advertisement used the trait of the medium in order to segment the audience and directly approach their target group.

When we add the component of segmenting the audience based upon choice of media, expressed through these examples, we can now add an additional criteria to use of the tool of Unification:

(1) making unconventional use of a component of the media;

(2) attempting to speak directly to a specific market segment by using the medium, its components or its traits creatively.

It is more difficult to implement this second aspect; however, attempting to do so can offer a focused and creative approach that will be remembered. Sometimes there is no need for market segmentation. When the message in the advertisement is directed at a wider audience, use the tool of Unification to express the message by any unconventional possibility the medium will allow.

Figure 1.8   Ullern Basketball Club needs new players: Take a brochure.

Other ways to use the medium

In a television commercial advertising English language courses at the Liceo Center, creative use was made of a component of the television medium. A Latin American chef stands in the kitchen of a restaurant, cutting vegetables. His boss comes charging in from the order counter and orders the chef to make a sandwich, but the cook just continues to peacefully cut the vegetables. The boss becomes incensed at being ignored. He yells angrily, peppering his speech with juicy curses – but to no avail. The chef doesn't even make the

Figure 1.9   Kitchen story.

slightest move towards making the sandwich. Throughout the advertisement, subtitles translate the boss's dialogue into Spanish. After a time, when the boss still doesn't calm down, the chef goes to the front of the kitchen – that is to say to the front of the screen – stands in front of the subtitles and reads them. Then the entire scene becomes clear – the chef only understands what's being asked of him when he reads the translation. He amiably returns to his station and prepares the sandwich (Figure 1.9).

The Dutch billboard campaign for Nurofen®, a product that eases headache pain uses a traditional medium: billboards. However, the advertiser chose to use the resources of the medium – the paper upon which it is typed – in a creative way. The paper itself is torn, actually splitting the head in the photograph and visually expressing the headache (Figure 1.10). This advertisement is a particular victory because pain is one of the most difficult conditions to represent. The advertisement introduces an unconventional use of the medium itself and gives the medium presence. It has gone from being a platform that allows us to express a message to a central and active element in the design of the message itself.

Differentiating a campaign in different media

Advertisers are often stuck in the belief that an entire campaign must be identical throughout all media. Had the Nurofen® advertiser been stuck on that determination, the previous advertisement would never have seen the

Figure 1.10 Splitting headache?

Figure 1.11 Don't just hope you have good insurance.

light of day; this particular Nurofen® advertisement could not be shown on the radio, on television or even in the newspaper. The billboard ad makes use of the specific traits of the billboard medium.

Each medium has different traits that can be used to deliver the message. This advertisement for the Fredrick Rauh insurance company makes use of the medium's trait of being flammable (Figure 1.11). The ad could not be duplicated into other media. Although using certain traits, like the trait of flammability, can be expensive, such advertisements create an exceptional effect so there is no need to hang one hundred posters in order for the message to be seen and remembered; a small number will do the job.

There is a defined thinking process to using this approach to the tool. The different stages are presented below.

INSTRUCTIONS FOR USE

(1) Precisely define the promise or the message.
The message is: London needs blood in order to save lives.

(2) Make a list of possible media (or of the media the client is interested in using). Create a list of resources for each media. Resources are the physical components of the medium and the traits of the medium. Traits can be (place, color, size, shape, etc.).
Although in actually doing this exercise, we would evaluate many media, we will choose one for the sake of the example.
We check possible media for the advertisement: a bus
Some of the resources of the bus are:
- windows;
- doors;
- color;
- shape – two storeys;
- place – the streets of London;
- movement.

(3) Search the traditional area in which each resource operates and try to define a new task for each resource – one that can support the message.
Windows: Point to bus passengers who could need your blood, or that you, God forbid may need their blood.
Doors: Every time the doors open an accident happens in the streets of London and another portion of blood is needed.
Color: Red, the color of blood.
The two-storeyed shape: Your body can carry or support the body of another person, who will use your blood.
Place in the streets of London: Point at the different houses whose residents need, or may need your blood.
Movement: for every mile the bus travels, another liter of blood is donated and saves someone's life

(4) Choose the resource that you believe has a new task which will deliver the message in the most convincing way. Create the advertisement by stressing the new task.

Figure 1.12   London's running out of blood.

The red color of the bus can help in the design of the message. If the color red symbolizes blood, and the message describes the lack of blood, part of the color will be taken off the front of the bus (Figure 1.12).

Approach 2: Content creates form: Recruiting a medium in the environment of the message

Choosing the medium because of the message

The thinking process for the examples we've looked at until now has involved selecting a medium and seeing how that medium can be altered or induced into delivering your advertising message. Another creative way for creating a link between the medium and the message is managed in the opposite way. In this method of Unification, the advertiser's choice of message and his evaluation of the components of that message will lead to making an exact choice of the most effective medium for delivering the message.

In Figure 1.13, an advertisement for a Land Rover jeep has been positioned on a poster near the top of an escalator. The sign reads: "You are about to discover what it feels like to use Hill Descent Control." The message is, of course, about smooth, stable and comfortable travel, even in difficult outdoor

Figure 1.13 Escalator warning.

conditions. The advertisers first decided upon their message. Then, physical sites were selected that personified the message with their own even, smooth movement. They then selected the medium of their advertisement to be in close proximity to the object that symbolized the even ride. The proximity of the sign to this element provided an actual physical trial of the chosen message.

Creating new media

A mental process of this type can expand the borders of the concept "media," allowing advertisers to select locations for delivering their message that may surprise the customers. This element of surprise can be successful in drawing greater attention. If we choose the medium after we define a specific message and consciously connect the two, we can produce unconventional advertising processes.

So, for example, the British ad agency Abbot Mead Vickers (AMV) BBDO chose to place advertisements against driving under the influence of alcohol in a site that a person who drinks alcohol would most likely visit: the bathrooms of the various pubs in London. Pub bathrooms are being used more and more regularly as a site for advertising because advertisers are realizing that people using the toilets are a "captive audience"; that they cannot avoid seeing – and reading – the ads. In the absence of other cognitive possibilities, bathroom

(a)   (b)

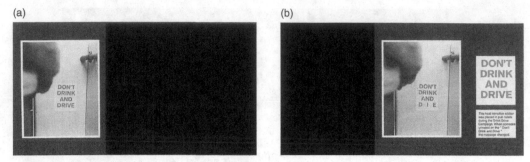

Figure 1.14   (a) Don't drink and drive! (b) Don't drink and die!

visitors may even welcome the diversion of the bathroom advertisements. The choice of the medium for any given advertisement will be made based upon the message itself. Here, the message discusses the dangers of too much drinking (Figure 1.14a). One of the physical locations immediately identified as a central component of this message is the urinal.

Perhaps simply placing the ad in the bathrooms would have created a meaningful effect with regard to its exposure to the relevant audience. However, choosing a specific, targeted physical location for the advertisement was just one of the unconventional choices that the agency made. After selecting the medium for the message by considering the message, the advertising team made a deeper connection between the message and an existing component in the environment of the medium. One of the components of a bathroom is also directly connected to drinking: the client's pee itself; and so they placed the advertisement on a sticker, inside the urinals. The moment the visitor "does his thing", the pee hitting the sticker will cause a chemical reaction to occur and change the text on it (Figure 1.14b). The result is that the advertisement is effective in every way possible: it is placed in a place that forces continued exposure and it connects between the medium and the message in the most immediate way, so that it becomes relevant to all people who see it.

Another example, surprising in its simplicity, is an advertisement encouraging tourism to a national park in Auckland, Australia. The message chosen was to stress its beauty. A central component in the message is, of course, the view from the site. While seeking an appropriate advertising medium, the advertisers asked themselves: why not use the views of the site itself to present the message? In effect, they chose the message itself as its own medium for delivering the message. The advertiser chose an out-of-the-ordinary tactic that enlisted park visitors in conveying the message to their friends and

Figure 1.15   Natural masterpiece.

acquaintances: they placed huge, golden picture frames throughout the park, in viewpoints and outstanding vistas. Park visitors felt naturally drawn towards taking pictures of the view and of themselves inside the ornate picture frames (Figure 1.15). When they then showed these pictures to their friends – they were "advertising" the park – showing off its natural beauty. The advertisers who invented this award-winning unconventional concept enlisted a new medium – people's own pictures – and connected it to the message – the park's natural beauty.

A systematic thinking process can assist in implementing this second version of the tool of Unification. Using this process, the advertiser can select a medium that will most strongly reflect the message and support it.

INSTRUCTIONS FOR USE

(1) Specifically define the message.

The message is: opposition to atomic energy testing.

(2) Make a list of the components of the message.

Think of a number of components that are associatively linked to this message:

- mushroom cloud;
- missiles;
- the button that activates the bomb;
- cancer.

(3) Review the components of the message and for each component, define
a possible medium that easily connects to the message. Try to select a
medium that will offer an immediate, visceral connection and will be in an
environment that will reach your target audience.

In this example, the advertisement is aimed at all citizens, not a specific
segment. For this reason, the medium has to have the most possible exposure.

Mushroom: Shelves of mushrooms in markets and supermarkets, an open
field where mushrooms are picked.

Missiles: Kid's rides.

Button: An elevator button, the traffic button you press to cross the street,
a clothing button, emergency buttons.

Cancer: Clinics and hospitals, places where you sell crabs.

(4) Choose the medium that is closest to portraying the element of the
message and is most accessible to the target market segment; create an
advertisement that combines the medium and the message.

The final campaign creates a bond between a component of the message, the
red button, to one of the media identified: the buttons that you press to cross a
street (Figure 1.16). The medium that was chosen powerfully brings the
message to life. The message is a warning about the accessibility of atomic
energy and how easy it will be to create a disaster. Every single person who
presses the button to cross the street, or even thinks of pressing the button,
gets the message.

Figure 1.16   The world could end this easily.

How can we benefit from resources in the environment of the medium?

Making use of external resources

In most of the previous examples we used the "internal resources" of the medium. Internal resources are any resources that we as "creators" of the advertisement can control – they are part of the medium. The following examples will use resources that are external to the medium. External components are components that have direct contact with the medium but are not part of the medium itself. If the doors of the bus, its windows and its colors are internal components of the bus, the road, the cars that pass nearby and the sky above it are its external resources. These components surround the medium and exist in the background of our sensory perception of that medium. We tend to overlook these external elements as potential resources when designing the advertisement. More precisely, we ignore the active task they could fill in delivering the message.

Unconventional use of the medium itself in a way that makes the medium present to the viewer, transforming it from a passive tool into an active participant in the ad, has been seen to create more memorable advertisements. Will recruiting a medium's external resources provide the same effect?

We already have an example that responds to this. In the advertisement against drunk driving, the advertisers made use of an element that was not in their control – the pee of the urinal visitors. The advertisement was only complete when this external element came into play. In fact, the creative use of this external element is one of the aspects that makes this advertisement so unique.

In order to take full advantage of possible external elements that can be recruited for use in delivering a message, the space surrounding the medium must be optimally surveyed. We can seek components from the environment to serve our advertisement in the same two ways we looked for internal elements: we can either look at components found in the environment of the medium or we can look for a medium that is in the environment of the advertisement's message.

Reducing the resolution of the medium: drafting an environmental resource

Earlier, we presented an advertisement for a local paper that pointed to different objects around the billboard upon which it appeared, in a way that

(a)

(b)

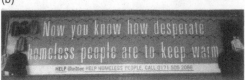

Figure 1.17   Now you know how desperate homeless people are to keep warm.

expressed the involvement of the paper in the life of the community – down to the smallest of details. Advertisements that use components of the advertisement's environment can be thought of as creative specifically because they do not recruit elements exterior to the surrounding of the advertisement, but use elements that are recognized and are more regularly seen by the viewer. Additionally, they stress the chosen message, while simultaneously acting as an immediate expression of it.

The next example (Figure 1.17) is not just a clever and exciting advertisement, but also serves as a perfect illustration of the principle of using the surroundings of a medium. A social campaign, the goal of this advertisement is to raise money for an organization that helps the homeless. The homeless, who live in the streets, are, of course, a component found in the environment of the chosen medium of billboards, if placed in the "right area" of the city. Pointing out the homeless in the environment of the advertisement has to be done with sensitivity in order to treat the homeless with dignity. Similarly, this sensitivity will help the advertisement avoid appearing self-righteous or over-demanding. The chosen campaign was based on hanging clothes on a billboard that is quite a distance from the ground. Over time, the clothes were gradually taken from the hooks and the message beneath was exposed. The homeless, the environmental component chosen, were used without actually having to be seen. Another advantage of the advertisement is that it transforms from a teaser to an ad that delivers the message without the need to actually change signs. The creative use of the medium generates the change itself.

INSTRUCTIONS FOR USE

(1) Compose the message as precisely as possible and choose a medium.
The advertisement is for a vacation spot in Palm Springs. The chosen message is "change the weather" and the medium is billboards.

(2) Create a list of the components of the medium's environment:
- land;
- road;
- shrubbery;

Figure 1.18   Try this.

- buildings;
- sky;
- lights.

(3) Test how each of the environmental components can fill the task of building the message.

Land: Replace the land with sand.

Road: The road can lead you to Palm Springs.

Shrubs: The grass in Palm Springs is greener.

Buildings: Enough being inside buildings, go out to the beach.

Sky: Exchange the cloudy sky with the blue clear skies of Palm Springs.

Lights: Exchange neon lights for the sun.

(4) Choose the component you want to use, by choosing which can play the most relevant and effective task in communicating the message of the ad when it is used in the new task you have assigned it.

The component that was chosen was the gray cloudy skies, which stresses through its opposite, the sun shining and the palm trees in the Palm Springs vacation spot (Figure 1.18). The advertisement does not sell an illusion, but supports itself by pointing to a known and given reality. Through stressing this, it intimates the possible alternative to it.

And some for the road

(1) Creative advertisements take into account the possibilities a given medium provides for audience exposure. At the same time, they deviate from

Figure 1.19   Scream Sunday.

the constraints of a medium in order to stand out. The deviation will necessarily occur through unconventional use of the medium itself.

(2) The tool of Unification allows advertisers to systematically review possibilities within the medium itself. By giving an existing resource a new task, you express the chosen message.

(3) When using the tool of Unification, the advertiser thinks systematically about using an internal resource of the medium in a new way, or about giving a new task to a resource in the environment of the medium in a way that serves the message and renews the medium itself.

(4) The effect of being amusing, witty – and remembered – is a free bonus when using the tool of Unification. No need to add expensive elements or to invest in additional media time. The use of existing resources does the trick.

(5) Using the tool of Unification can improve the creativity of an advertisement and the length of time it will be remembered. Often, when use is made of a novel medium, the customer need only come in contact with it briefly in order to remember the message even after the campaign is over. Note this fact the next time you stand at the edge of an escalator, when you visit the men's room or, as in this bloodcurdling advertisement (Figure 1.19) – when you walk by a water fountain.

2 Activation

Gluttons for attention in a noisy world

We live in a sea of advertising. Potential clients literally drown in thousands of advertisements that come at them from all sides, at every moment of their day: while watching television, checking email, reading the paper, driving the car, listening to the radio, watching a movie, or simply looking out the window. Advertisers and marketing managers are constantly challenged with trying to figure out how to differentiate *their* commercial so that it will stand out in this hectic environment and grab the attention of the target audience.

Target audiences don't actually relate to most of the promises made by commercials as truthful or reliable. Messages like enjoyment, fun, personal fulfillment and effectiveness are so common that linking them to consumer goods just seems like a marketing trick.

So, in this messy mixture of empty promises and loud advertising voices, how can an advertiser get his target audience to seriously consider the message of his advertisement? How can he get the man on the street to invest time, energy and attention in "getting" the message he is trying to convey with his advertisement?

The message is tuned out before it is spoken

We are all familiar with commercials that encourage us to donate money or time for important causes: fighting cancer, helping the homeless, providing for kids suffering from developmental difficulties. Although we, as viewers, know how important these causes are, we tend to glaze over these kinds of ads. There are a number of reasons for this. For starters, there are so many of them. The sheer number makes grasping so many messages difficult, if not

Figure 2.1   AIDS research. Caption reads "To kill the aids virus you must put your finger on this letter."

impossible – so we let them pass us by. Even with the one message, it is often difficult to devote time to negative thoughts or unpleasant messages in the rush of day to day activities. Secondly, the disturbing aspects that are raised in these types of ads – like information that induces fear or pictures that are unpleasant to look at – again, lead us to faze them out. The award-winning ad (shown in Figure 2.1), like many of its kind, is trying to get the reader to make a donation to AIDS research.

Did you put your finger on the "s"? If you did, you may understand why this advertisement is more likely to succeed in drawing attention. More to the point, it probably achieved its purpose more effectively than more conventional ads. The answer to the "How" question lies in the fact that unlike most ads which require the reader to just passively read in order to get the message, this type of ad encourages the active, physical involvement of the person looking at it. By means of a slight manipulation activated on the viewer, he becomes a real partner in the creation of the advertisement's message.

Defining the tool

Notice that this advertisement can be described as a product of the tool of Unification, because it makes unusual use of one of the internal elements of the medium: the letters that make up the copy are used in a way that stresses the message itself. However, this advertisement has a unique characterization that differentiates it from the tool of Unification: it demands active involvement, actual physical involvement of the viewer in order to appreciate the full effect of the message.

As we saw in the above example, Activation encourages the viewer's involvement by requesting his physical trial of something in the advertisement itself. As opposed to the usual process, in which the advertisement itself spells out the message and the reader simply reads it, the process of Activation invites the customer to actually do something and through this activity, come to some sort of conclusion. The viewer himself is an active element in his own recognition of the relevance of the message.

Advertisements that use Activation require the viewer to make an immediate, physical action when standing in front of the advertisement in order to release the advertisement's message.

Other examples that prove the rule

The Activation in the advertisement shown in Figure 2.2, which advertises the broadcast of the Wimbledon tournament on a Pay TV channel, works in

Figure 2.2   Wimbledon.

a different way than the previous example. Although it does not seem to require a physical action of the viewer, it demands that the viewer read the text in a way that approximates the experience of watching tennis. By moving his head from side to side the reader becomes an active part of the advertisement, and through this movement, the message is conveyed. Thus, we can fine-tune the definition of Activation by explaining that when this tool is used in an advertisement, the advertisement makes a passive or active request of the viewer to actively do something – but not necessarily an active bodily movement – in order to get the message. In this case, the way the text is set up forces the reader to read in a different way. The message can not be ascertained without active viewer participation.

You can't ignore what you create

This postcard (Figure 2.3a), warning teenagers about the damage caused by exposure to the sun was handed out at the beach and at youth hangouts. The challenge for the advertiser was clear: faced with the temptation of having luxurious suntanned skin today, young people tend to be apathetic to the

(a)

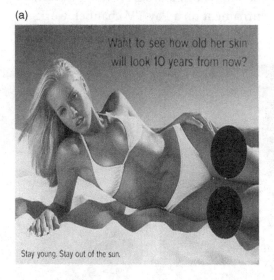

(b)

A: Ammirati Puris Lintas, Sydney ⊂>: Dean Mortensen ⊂>: Matt Baldwin 🖾: Alister Clarke

Figure 2.3   Beach babe.

"future" dangers of skin cancer, wrinkles and skin discoloration. Delivered in a conventional way, there is clearly the potential for such a message to be perceived as lecturing and annoying making the chances it would remain unread high. But this next advertisement does not allow its intended audience to ignore it (see Figure 2.3b). The postcard asks a question, perhaps even a banal one, which, if appearing as a headline, may have a minimal or non-existent effect. However, in this ad, in order to answer the question, the reader is invited to perform a physical action that will be hard to resist: put two fingers into the holes. The effect of doing this is intense – and sufficiently disturbing – to jolt the reader into understanding the severity of the message. The effect is immediate too – the viewer has demonstrated to herself the consequences of ignoring the message. More to the point, the viewer herself played a central role in creating the message. In a subtle twist, the message is no longer coming from some third party, but from the reader herself. When the viewer is turned into an active participant in creating the image, the message will be that much harder to ignore.

When is Activation useful

Activation invites the prospect to make an immediate, physical action during the encounter with the advertisement by using the components of the medium itself or the viewer's body. The aim of this action is for the viewer to reach the conclusions at which the advertisement aims. The special characteristics of this tool make it particularly useful for conveying two types of message.

(1) *Making the client aware of a problem.*
(2) *Making the client aware of a solution or benefit offered by the advertised product or service.*

Making the client aware of a problem

The Activation tool is a very effective way of making the client become aware of a problem that the advertiser offers to solve with the product or service. Often, clients aren't aware that they have a certain problem, or if they are aware, have learned to put it out of their mind, not seeking immediate solutions. Occasionally, products solve problems of which the client would prefer not to be reminded.

Another example is an award-winning commercial for an anti-dandruff product. This product is a prime example of a group of merchandise that poses a specific difficulty for advertisers: they suggest problems with which potential clients prefer not to identify. Potential clients do not want to see themselves as suffering from issues such as excess weight, sexual infertility or hair loss. They generally don't want to be reminded of the personal relevance of products for problems of this nature. These products are therefore harder to advertise.

When designing an advertisement for products like these, the advertiser's first tactic is to choose a discreet and inoffensive manner to convey the message. In this way, the advertiser can avoid pushing the client away or raising his antagonism. At the same time, the commercial must make the problem "real" in the mind of the consumer, so that he will invest time in familiarizing himself with the solution offered in the advertisement and will be convinced of its necessity.

The anti-dandruff advertisement was placed in a double spread. The left-hand side is pitch black with a text saying: "Shake your hair over the left side to see if you should read the right side of this ad." The text on the right-hand side of the ad, listing the advantages of the anti-dandruff product would seem to break every rule in the proverbial advertising handbook: it is long, it is very wordy and it demands that the viewer invest a great amount of time and mental energy in reading it. If the text had appeared on its own, it would be reasonable to assume that the advertisement's effectiveness would be close to zero. One would imagine that few would take the time to read it. But the advertisement employs a manipulation that is difficult to resist: the offer of a discreet, personal, non-threatening dandruff check. After the potential customer discreetly identifies his own problem in a way that he believes to be reliable and upfront – he did the test himself and saw the results with his own eyes – there is a greater chance that he will invest time in reading the detailed advertising text on the right-hand side of the page. Even a person who does not suffer from dandruff will see white specks appear on the black paper when he passes his hand through his hair and therefore may be inclined to read on.

By means of a test, the advertisement induced the potential customer to recognize a problem of which he perhaps wouldn't have been previously aware, and at the same time suggests a solution. Activation is effectively employed in order to add emphasis and impact to the kinds of "sensitive" messages people are inclined, whether consciously or unconsciously, to avoid. It can be used for messages that stress an individual's weakness or negative quality, or for a danger or problem that may occur if he doesn't use the product or the service offered. In actual fact, it can be used to invoke any problem that a product is designed to solve.

Figure 2.4 shows an ad from a campaign for chiropractic centers treating people with neck and back problems. The advertisement was strategically

Figure 2.4   Does this hurt?

placed to make it awkward for the viewer to access it. To read the ad, the viewer had to bend or twist his head or back to the side.

By coaxing the viewer to read the ad in an awkward way – an effort not usually demanded by an advertisement – the ad both identifies its potential clients, raises awareness of the problem and offers the solution. Moreover, because the message is experienced rather than passively received, its impact is strengthened. Like many advertisements that use Activation as a tool, the manipulation created by the advertisement is discreet and witty; it is not invasive or insulting. It makes the viewer take personal ownership of the problem and presents a suggested solution.

In Figure 2.5, the advertiser used an optical illusion to represent the unpleasant problem of acne. Presenting the problem in another way may have been perceived as vulgar. By using this optical illusion, and the interest it creates, the ad helps overcome the viewers' initial reticence to engage with the subject.

Special qualities

As we have seen, Activation is an effective way of making potential customers aware of the need for certain product. It can often generate a sudden awareness in the client, such as "I need dandruff shampoo" or "I've got to do something about my back".

However, when an advertiser decides to use Activation to make the problem apparent to his customer, he must be sure that the advertisement is

As soon as a black spot appears, use Clearasil.

Figure 2.5   As soon as a black spot appears, use Clearasil.

not intended to promote a specific brand or product. The audience will primarily remember this type of advertisement as promoting the category but not necessarily the brand. For the non-brand leader this may prove a weakness, especially in a market full of similar products fulfilling the same need. Activation itself does not create a connection to a specific brand. This means one must create that connection through other means, such as fusion of the brand with a message, an action that will be discussed with regard to other tools later. Activation is best used when the brand is immediately identified with the need, and when that identification is as exclusive as possible.

Making the client aware of a solution

When using the tool of Activation to deliver the second type of message, making the solution apparent to the prospect, the message is also emphasized through a physical activity that the advertisement invites the viewer to perform. Through this activity, the potential customer can experience first hand the outstanding qualities of the product or service and/or appreciate the benefit that comes from the feature being promoted.

The following print ad uses an optical trick to advertise a cleaning product. The text that appears at the bottom of the ad asks the reader to focus their gaze on the ACE logo; a few seconds later, the stain that surrounds it miraculously disappears (see Figure 2.6). Participation in the game reveals the message. The

Figure 2.6   The ACE logo.

(a)

(b)

Figure 2.7   Get the door as you go.

activation here serves the message directly and supplies a reward: those who take the time to try out the ad's suggestion are surprised and interested – they've had fun.

The example of the Swedoor® company (Figure 2.7), a Swedish door manufacturer, is perhaps a more clumsy activation, but it is amazingly effective in making the benefit present. The company actually handed out life-sized posters of their doors for clients to take home and size up.

Activating imaginations

Activation ads need not be expensive. Nor must they use the medium in a unique way, as this previous example and several others have. The ads shown (Figure 2.8) use the tool of Activation in quite a different manner.

(a)   (b)

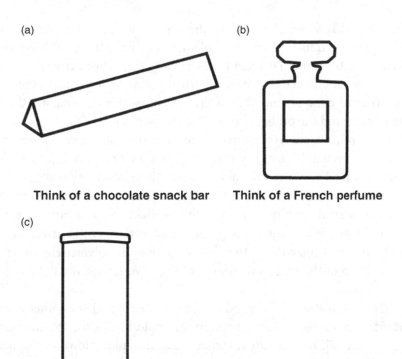

Think of a chocolate snack bar   **Think of a French perfume**

(c)

Think of a potato chips brand

Figure 2.8   (a) Toblerone®; (b) Chanel®; (c) Pringles®.

But why are these examples considered an Activation? We've maintained that to classify an advertisement as using the tool of Activation it must require the viewer to perform an immediate, physical act through which the message of the advertisement will be conveyed. It is absolutely clear that the advertiser does not think that anyone reading the ad will actually do something – besides imagining. Yet, this advertisement is still considered an Activation.

This advertisement is an example of a specific type of Activation called "Mental Activation." It asks the viewer to perform a specific act, which the viewer will not actively do. Nevertheless, the process of thinking through the action creates an effect close to having performed the action itself. Therefore, it fits into the category of Activation. So, even if the client does not actually do the activity, but the advertisement asks him to do something in a way that will make him do the activity in his mind, thus giving him a similar reaction to actually implementing the activity, it is considered a "mental activation." The client undertakes a cognitive activity that deviates from the general activity required when taking in a regular advertisement. In order to comprehend the advertisement, you must imagine the activity demanded and the probable results of that activity.

However, an advertisement of the type that promises "if you buy this product you will be a happy person" is not a mental activation. It does not invite the reader to perform an immediate physical activity and does not cause the client to feel the result of the imagined action. In addition to that, the actual physical activity of buying the product will not bring the result of being happy. When the message of the advertisement is proffered directly and without connection to the viewer's action, the advertisement will not fit the criteria of the tool Activation and the advertisement will be considered less creative.

This ad for the Jeep® Grand Cherokee (Figure 2.9) is another example of Mental Activation. As in the previous example, it is reasonable to assume that the viewer will not actually do what is asked of him. However, even without actually doing the activity, the request in the ad will give rise to the rough feeling of tongue touching paper and it is this feeling that is integral to the message of advanced traction. The Mental Activation in this case creates a contrived feeling that directly serves the message.

Benefits of using Activation: What it provides

Activation provides certain advantages in delivering a message. Instead of reading, watching or listening to the message, as is customary in the interface between the advertisement and the target audience, the viewer is required to take an active part. This activity will bring the message to life for him. The viewer's personal involvement, triggered as a result of the activity, reinforces the impression that the message leaves. Because the message is not reported by an external source that can be suspected of unreliability or non-objectivity, the message is experienced impartially. The viewer is invited to take part in an immediate, concrete and physical test, which is a deviation from what is usually expected of him when viewing an advertisement. This focuses attention on the specific problem and on the possibility of its being solved. Or it highlights the uniqueness of the specific product feature that comes to light in the message. The physical activity also offers additional

Figure 2.9   Stick your tongue here.

advantages. The time the viewer spends on the advertisement increases dramatically, which positively affects the degree to which the viewer will remember the advertisement. The overall experience will also be more profound than that experienced with a regular advertisement.

How to apply Activation

(1) Compose the message as precisely as possible.

The advertisement is part of a road safety campaign.

The message we will choose for the campaign: speeding near a school crossing can cause tragic results.

(2) Plan a scenario that involves some kind of test, requiring activity of the relevant senses; the result of the test should prove the truth of the message.

The target audience will see the results of speeding near a school crossing. Through this test, viewers will be faced with the reality of what their actions can cause.

(3) Choose the medium that is most appropriate for the test (newspaper, television, radio, billboard, etc).

In this case, as we are using visual images, any potential direct activity must take place when the viewer is in front of the advertisement. We will choose a newspaper as the medium.

(4) Think of the way that the target audience can physically participate in the scenario, while using the chosen medium.

(5) Find the most appropriate and convincing way to invite the target audience to do the trial. This invitation must serve a few purposes: (a) it must point to the potential benefit accrued from the trial; (b) it must provide simple, clear instructions for a reasonable trial; and (c) it must be an appeal to the natural curiosity of the target audience.

We can, of course, show images of a child before and after a car accident, but this, and other ideas of this type, won't activate the viewer. Instead, we can present a picture of a happy child and an activity in which the viewer is capable of causing him harm (see Figure 2.10). If the viewer follows the instructions of cutting the dotted line, he comes face to face with the damage that he can cause. The action serves the message in that viewers, most likely, will not want to follow this activation, since they wouldn't dream of hurting a child, even a picture of one.

Just rewards

An important aspect in planning the test is the amount of investment the viewer is asked to invest, in relation to the reward he will receive from the trial. An advertisement that uses Activation but does not give sufficient

(a)

(b)

Figure 2.10   (a) Open here; (b) Tear out here.

reward – like a new understanding of yourself, a surprise or just a smile – can create negative feelings. Viewers can feel a bit cheated or taken advantage of after trying very hard with no reward.

The next example (see Figure 2.11) is an Activation that rigorously follows all the rules discussed: it is interesting, immediate and asks for physical action. The action required – to connect the dots – also serves the messages in a direct way. The weak link, however, is in the quality of the reward the viewer receives for his effort in connecting the dots. All he sees is the company's logo, which is quite a disappointing solution. The Activation implemented in the advertisement could have been more effective if it had offered a surprising or more creative reward that would justify the effort required.

Planning the test

In planning the test we must make sure that the benefit received is higher or equal in value to the effort invested. At the same time, we must make sure that the viewer has all the resources necessary to accomplish the test at

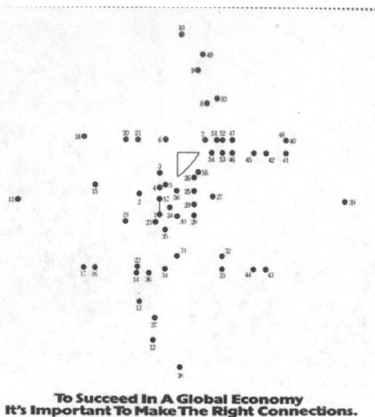

To Succeed In A Global Economy
It's Important To Make The Right Connections.

To find out what's new at N.E.T., begin by placing your pencil on the number one. Ah, but it's far more than a new logo. It's a broadened vision of enterprise networking. One that places N.E.T. squarely at the forefront of the networking revolution. With products and technologies that will connect you desktop to desktop, down the hall and around the world.

NETWORK EQUIPMENT TECHNOLOGIES

Figure 2.11   Making the right connections

the moment that he happens upon the advertisement. Therefore, advertisements that involve the use of a telephone, cutting out of coupons, or sending a postcard in the mail, are not considered as using the tool Activation, even though they involve a physical activity on the part of the client. This perhaps, is another drawback of the previous example, as it required the use of a pencil.

The elegance of the tool and its creativity are expressed in the fact that the Activation is performed through the exclusive use of the advertisement, the medium in which it appears and the client himself. The elements all belong to the Closed World of the advertisement itself and provide all that is needed for the implementation of the Activation.

Taking advantage of the medium

The following examples illustrate the different ways a viewer can be activated when only using the Closed World of the newspaper. In this case, the Activation generates an activity that, when the viewer performs it, forces him to directly confront the message. In the following two advertisements, for example, the trial is based upon folding the pages or splitting them, in order to convey the message (see Figure 2.12). In this instance, if the final image would appear as it is, the advertisement would be vulgar and off-putting; but when the viewers create the image themselves (as we saw in the anti-cancer postcard), the message is received more effectively.

The next example of Activation remains in the Closed World of the interface between the client and the advertisement. The ad was placed in the sexual services section of the personals – and looks just like any other sexual service advertisement (see Figure 2.13). Nothing in the ad itself invites suspicion that the message of the ad is actually completely different. Pamela is apparently curvy, blond, and attractive. However, when Pamela's potential customers call the telephone number provided, they hear a recorded message. A soft and seductive woman's voice apologizes to the now frustrated customers for not being able to provide the service. The imaginary Pamela is unable to service her clients, she continues in

(a)

(b)

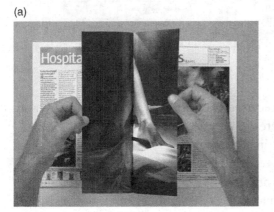 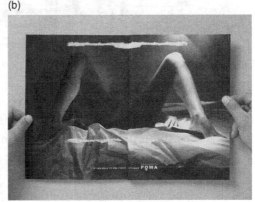

Figure 2.12   Force and effect. Caption reads "If you have to use force it's rape."

Figure 2.13 Phone for sex.

the recording, because she has received the results of her AIDS test – and it was positive. "If you are interested in pleasant, safe, and healthy sex," continues the recording, "I am living proof that you must always use a condom." This Activation takes advantage of an action that the client would generally do when faced with this sort of advertisement – phoning – and therefore the use of the external tool (the telephone) still leaves us within the Closed World of the advertisement.

Unlike easy-to-forget advertisements that encourage the use of condoms, this advertisement directs itself exclusively to the specific target audience in the high-risk group. The manipulation implemented by the advertisement borders on real fraud, but because of the discreet nature of the test and the surprise element, the advertisement is perceived as creative, effective, and specific, in that it makes a direct entreaty to a very specific market segment.

All the examples make a new use of the medium itself. The manipulation that is created through this new use serves the chosen message. There are, therefore, many parallels between the tool of Unification and the Activation tool. Unification, as was presented earlier, uses the medium itself in order to strengthen the message. Because Activation is based upon manipulation occurring inside the Closed World of the advertisement and often makes new use of the medium in order to make room for the test to happen, certain characteristics of Unification are present. Overlapping between the two tools increases the opportunity to arrive at a creative idea.

The ideas presented in the advertisements shown in Figure 2.14, for example, could have been reached through the use of Unification and/or

(a)

(b)

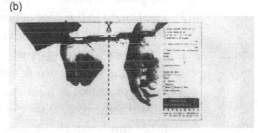

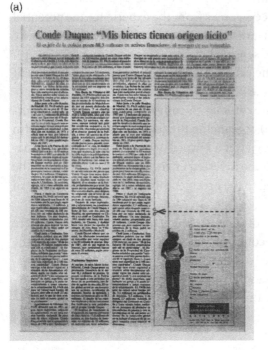

Figure 2.14   Let loose.

through Activation. The advertisement's aim is to recruit volunteers, supporters, and donors to Amnesty. It, therefore, includes a short form to fill out and send to the organization. Calling upon the Unification tool, one would ask oneself how to use the components of the advertisement to serve Amnesty's message. One of the components is the dotted line that marks the place to cut the form; it is possible to use this line as the borderline that divides between freedom and slavery or between life and death. The visual could have been used simply to create separation through use of the dotted line – between images of freedom to those of slavery. However, the creators of the advertisement intensified the effectiveness of the message by employing the tool of Activation. Note what they induce the viewer to do mentally, or actually, if he cuts off the form.

And some for the road

(1) The tool of Activation is based on two central values: manipulation and test. The advertisements implement a manipulation on the target audience in order to bring them to active participation in the test or action suggested in the advertisement.

(2) The benefit in this type of process comes from the experience the potential customer undergoes. The advertisement provides him with a physical experience involving dynamic sensory activity. This activity raises awareness to the specific problem the product would like to solve or to a benefit that is promised as a result of using the product.

(3) Advertisements of this type are difficult to compose because they undertake a three-part challenge: (a) convincing the target audience to perform a certain action; (b) giving the viewer a reward or benefit for his participation that is greater or equal to the invested energy; and (c) expressing the problem or solution connected to the product through the performance of the act itself.

For this reason the number of advertisements based on this tool are relatively few. However, the commercials that succeed in following the required steps will be perceived as being especially creative. More importantly, they will succeed in leaving their mark on the prospect's consciousness in a more profound way than the majority of passive advertising.

3 Metaphor

Splitting metaphoric hairs

Advertisers often seek methods to renew the way a consumer perceives a product. One method used to accomplish this is metaphor – or creating a manipulation by using a symbol. Metaphor is the most used – and abused – tool in advertising.

A happy family appearing in an advertisement for a family car is a common sight in advertising. The symbol of the "happy family" tells us something about the family car and will hopefully make us see the car in a new light. Sometimes advertisers will use a medal to signify one product's superiority over others. These metaphors are generally considered banal and not frequently found among award-winning advertising.

However, on occasion, the use of symbols to deliver an advertising message can be considered more creative. The following two metaphor-based advertisements have won awards, for example. The two ads in Figures 3.1 and 3.2 are completely different from one another. The first warns against the dangers of smoking and the second is an advertisement for a jazz festival that will take place in an unusual place – the Minnesota Zoo; but both are expressions of this "other" form of metaphor.

Why do these two representations "work"? Why are they judged to be "creative" while the "happy family" is considered trite and commonplace?

Getting it in one shot

One of the reasons is that these ads employ symbols that are connected to the message of the advertisement. In both, there is a physical fusion between the symbol and the product or service being promoted. In

Figure 3.1   Cigarette end.

the first advertisement, the symbol of the human body in the shroud clearly expresses the message of death. The visual association between the body in a shroud and the cigarette makes the need for explanatory text redundant. The message is clear simply by looking at the image. In the second advertisement the structure is identical. The monkeys represent the message of the zoo and their physical similarity to musical notes links between the symbolic image and the product in the mind of the viewer. Looking at only one image, we see music in the zoo.

The creators of these two ads give all the necessary information in one image. This is in contrast to the example of the happy family and the car, in which the viewer must look back between one image and the other to get the message. In award-winning metaphorical advertisements, the message, the product and the symbol are all fused into one very clear – and often thought provoking or interesting – visual representation.

The Metaphor tool

Thus, it is possible to distinguish a very particular use of metaphor that works as the basis for creative advertisement. We call this tool Metaphor.

Figure 3.2   Jazz at the Minnesota Zoo.

Ads created by using the Metaphor tool are characterized by performing a trick or manipulation on a recognized and accepted cultural symbol that immediately connects it to (but not in an obvious manner) the desired message. The symbol in each of these examples does not appear by itself: it is fused with the product, such that the visual presented in the advertisement contains two objects that are perceived at the same time: the product itself and the representation – or the symbol.

Use of such a symbolic system in advertising can offer the advertiser a double benefit. On the one hand, an advertisement based on a symbolic system will be sophisticated and thus require the viewer's interpretation. Because of this, a viewer will necessarily spend more time on the advertisement. On

Figure 3.3   Security assured.

the other hand, symbolic advertising creates an immediate cognitive effect. Because the advertisement relies on a visually created connection between the symbol and the message, it is very effective. The information is delivered by the visual itself, requiring minimal explanatory text.

Fusion, a concept we have referred to several times in this chapter, is the act of creating a strong visual bond between the symbol and the product, so that they are perceived as one. The Durex® press ad in Figure 3.3 is an excellent example of this. The product itself is used to create the symbol – the lock. The condom itself is the lock. What stronger way could they find to tell that particular message?

What is a symbol?

In 1925, Victor Shklovsky, a leader in the Russian Formalist movement, published an influential treatise in the field of art criticism, called "Art as a Device." Shklovsky pioneered the characterization of art as an expression that uses representations – or symbols – manipulatively. According to Shklovsky, a symbol is any object that points to another object or concept in one of two possible ways.

(1) It replaces a particular object with the cognitive model that includes the object itself and others like it through the use of one image. Imagine an apple. The mental picture of the object allows you to compare it to other "apples" and to make decisions about what an apple is and what an apple is not.

(2) It connects the object's main trait to the main trait of a cultural symbol. So, therefore, the word seduction can be presented by using the object "apple."

Symbols operate on two parallel levels: on the cognitive level, they allow us to conserve mental storage needed for the recognition of objects. We see an apple and know this as "apple" like those we have seen in the past. We see an apple and understand "seduction."

The symbol serves to economize an individual's mental storage space, so that it is easier to deal with a world flooded with environmental stimuli. The symbol is a cognitive tool that simplifies the process of human information absorption, but at the same time, in its second – or cultural – task, symbols serve to complicate and cloud the direct assimilation of objects and concepts. When we use an apple to represent seduction we create a metaphorical–cultural connection that delays the immediate perception of the concept. We create an intermediary that inhibits assimilation of the concept. The artistic symbol, argues Shkolovsky, is a representation from the symbolic – cultural system and in this way is differentiated from the economic task of non-artistic symbolism. Using this distinction, Shkolovsky proposes criteria to distinguish between what is art and what is not art: any symbol that adds an obstacle to perception so that the individual will have to refocus attention onto the object belongs to the system of art.

Distinguishing art

Advertising uses symbols as one of its primary tools. Shklovsky distinguished symbols that are art as those that force viewers to take a step back and consider their perception of an object. We will use this distinction to discriminate between creative advertising and advertising that is not creative. When symbols are used in non-creative advertising, it does not make the viewer take a step back to re-evaluate his perception of the object or product itself. However, in creative advertising, the viewer's perception of the product is clouded, forcing him to see it in a different light than before and renew his perception of the product.

The process of recognizing consumer goods becomes automatic with time. As a product becomes common and recognized by the consumer, he will more easily recognize the product without devoting any attention to it. Advertising uses symbolism in order to renew the client's perception of a product. Shklovsky writes: "Art exists in order to recreate our perception of life. To make things felt. To return "stone-ness" to the stone. The goal of art is that the person should feel things, will perceive them with all his senses and not only recognize them." Isn't this the ultimate goal of advertising?

How to effectively create a metaphoric ad

The effectiveness of an advertisement using symbols is based on two factors: the intelligent choice of the symbol and maximum fusion between the product, the symbol and the message.

Choosing the symbol

The more closely connected a chosen symbol is to the chosen message, the more effective the advertisement will be. There is no need for market tests or focus groups in order to select effective symbols. The advertiser himself lives in the culture of the potential consumer and thus it is reasonable to assume that the symbols will be available to him. So, if the chosen message is "efficiency," for example, we can easily list a number of cultural symbols that will represent this value:

- a swiss clock;
- an ant;
- a super-computer;
- Japanese people;
- Germans;
- elves.

For the message "cleanliness" we can list the following symbols:
- snow;
- water flowing from a spring;
- the color white;
- transparency;
- lightning;
- surgical equipment;
- a pharmacy.

Different strokes for different folks

Symbols are, of course, culturally dependent and period dependent. Sometimes symbols change from place to place and from one population segment to the next. In evaluating symbols, advertisers must plant their considerations in a cultural context specific to the advertisement's target audience. Because of this, the above lists of symbols are only examples, as are the other lists in this chapter. The lists are not comprehensive or compulsory. Readers located in different cultures or sub-cultures may have various – and justified – reservations about any particular symbol on the list.

Clear symbols

The two examples shown in Figure 3.4 are ads for the sports product manufacturer, Penn®, a company most widely recognized as a manufacturer of tennis equipment, especially tennis balls. Penn® is also a well-known promoter of the sport. The advertisements draw attention to the company's sponsorship of tennis tournaments in Canada and France. The symbol chosen to represent France is a croissant and the Canadian symbol is a hockey puck. These two symbols are unambiguous. What they are meant to represent is clear and at the same time, they connect directly to the advertisement's message.

Creating a series of symbolic messages

When selecting symbols, an advertiser may decide that he would like to create a series of advertisements using more than one symbol for his campaign. When several symbols can suggest the same idea, it is usually possible to compose such a series, with each advertisement delivering the same message.

In the following example, Bally®, a company that makes shoes sized by width as well as length, chose the concept of freedom for the message of their advertising campaign. Possible symbols for freedom could be a tropical island, blue skies, the sea shore, a yacht at sea, a surf board, an open field or a forest, a bird, a butterfly, an airport, a fast motorcycle, and more.

(a)

(b)

Figure 3.4   (a) Croissant; and (b) Hockey puck.

Bally® decided on a campaign made up of four commercials (Figure 3.5). In each, a visual connection was made between one of the symbols of freedom and a component connected to the product: a foot.

Pitfalls to watch for

Although Bally®'s campaign commercials were recipients of creativity prizes and even though they deliver the chosen message clearly without requiring explanatory text, the advertisements still have shortcomings. Look at the placement of the product itself in the advertisement. Because the chosen symbol does not connect to the product itself but to an exterior component connected to the product (the foot), the customer may not connect the advertisement with the specific brand of Bally®. The shoe only appears marginally in the advertisement. The advertisement could have been more effective had it created a tighter visual connection with the shoe and not with the foot. The drawback of this campaign brings us to the second requirement for effective use of the Metaphor tool.

(a)   (b)

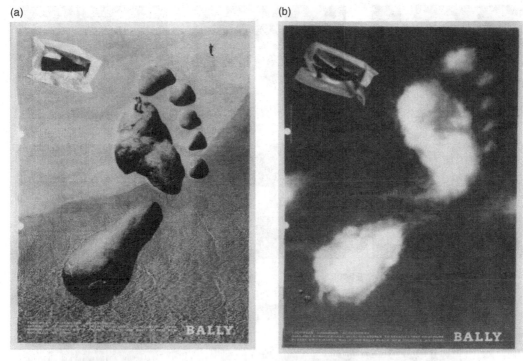

Figure 3.5   (a) Island; (b) Cloud.

Maximum fusion between the product and the symbol

Earlier examples, such as those of the Penn® company sponsoring tennis tournaments, perfectly meet this second condition. Ads like these create a visual connection (fusion) between the symbol and the product. The deeper the physical connection between the components of the symbol and the product itself or the central components of the product, the more easily this advertisement is likely to be remembered in the specific context of the product advertised. In these examples, a technique was used in which the product and the symbol were combined into each other. Components of the symbol (croissant, hockey puck) are in the color and texture of tennis balls, and display the Penn® company logo.

In a TV commercial from Bangkok designed to encourage milk drinking, the chosen message emphasizes the importance of milk's calcium for strengthening teeth. The commercial is minimalistic: a row of glasses set up in a semi-circle are slowly being filled with milk. In the first stage, when only some of the glasses are filled, the row looks like a mouth with some teeth missing. The voice-over announces: "That's what your teeth will look like if you

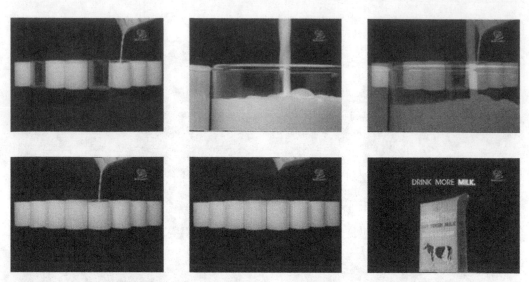

Figure 3.6   Milk is needed for strong teeth.

Figure 3.7   The walnut.

don't get enough calcium." After a short break, the other glasses are filled and the row looks like a line of straight, white teeth (Figure 3.6). The announcer proclaims: "For white, strong teeth, drink more milk."

Many advertisements that make use of symbolism don't make use of the full potential available in the Metaphor tool because they present the symbol next to the product, but don't create a more significant fusion. An example of this would be, as previously stated, the happy family near the luxurious car.

Other advertisements use a symbol that represents the chosen message, but do not create a fusion between the symbol and the product, as in the example from the Volvo® commercial (Figure 3.7).

Figure 3.8   Stop apartheid.

The advertisement for laundry fluid shown in Figure 3.8 creates a fusion with items in the Closed World of the product, but not with the product itself. Each product has elements that relate to its use (with laundry detergent, these could be a washing machine or laundry) and these elements are called integral elements in the product's "Closed World."

The advertisers wanted to offer the message that the product eliminates the need to wash whites and colors separately. To do this, they chose to represent the message by the concept of apartheid. Presenting the concept of apartheid in the unusual context of laundry, they created humor and eradicated the more threatening and upsetting elements of the concept. By using elements in the product's Closed World, they've ensured that the message will be remembered with regard to laundry. However, although this advertisement creates a successful – and funny – fusion between apartheid and laundry, it does not create a fusion with the product itself and, therefore, runs the same risk that the identification of the message with the product will not be complete.

An example: Options open to the advertiser

When choosing a symbol, we recommend that our clients complete the process through visual fusion between the symbol and the product. Many times, this tactic reaps more creative results, as defined by advertisements that win creativity awards.

The following process may clarify the importance of this fusion as a means for increasing the advertisement's effectiveness.

The chosen message is, of course, that this car is effective in avoiding skidding. The symbol chosen to communicate this message is a bath

(a)

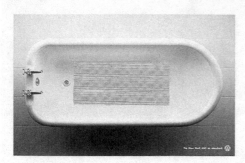

(b)

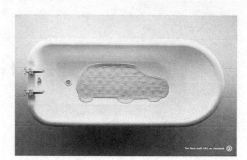

(c)

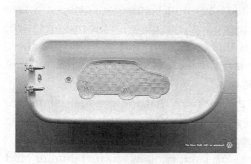

Figure 3.9   Non-slip model. Caption reads "The New Golf. ABS as standard."

mat – a common object, known to all, whose function is to prevent slipping (Figure 3.9a). But this advertisement does not create significant fusion between the symbol and the advertised product – a Volkswagen Golf. A possibility to create more significant fusion between the symbol and the product appears in the second example (Figure 3.9b).

In this part of the figure, although the message is fused with an element in the world of the product – the shape of a car – it is still not immediately identified with the Volkswagen Golf. A possibility for greater fusion – maybe too great – appears in the third part (see Figure 3.9c).

Going all the way – fuse all three

An advertisement using the Metaphor tool will be most effective when based on the bond created between the product, the chosen message and the symbol, as is clarified in the following diagram:

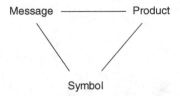

Advertisements that create significant fusion between all three elements in one visual image augment the connection between product, symbol and message in the mind of the viewer. Intuitively, it would seem that this process would necessarily increase "memorability" of the product and its message. It also has been proven that ads of this kind win creativity awards.

The act of fusion occurs when the advertiser creates a visual connection between the chosen message, the product, and the symbol. Successful fusion creates immediate visual connection between components of the product and components of the symbol. This can be accomplished by using certain graphic tactics. One possibility is to design the image of the symbol by exchanging one of its components with the texture of the product or with a graphic element identified with the product.

Butter is advertised in the example shown in Figure 3.10. The chosen message stresses the inclusion of sea salt in the butter. The symbol chosen to represent the sea salt is ocean waves and the visual fusion shows a wave in the texture and color of butter. The wave itself also bears the grooves of a butter knife.

Figure 3.10   Butter, made with sea salt.

Figure 3.11   A flipper that has grip.

The example that is shown in Figure 3.11 is an advertisement for car tires that provide smooth and safe movement in wet conditions. The symbol chosen is diving fins, fused with the product and the message by picturing fins that have the texture identified with tires. These two examples both show how all three elements – product, message and symbol are delivered in one visual image.

Sometimes the fusion occurs by using the product itself to replace one of the components of the symbol. A specific example is an advertisement for spicy French fries. The chosen message is the spiciness of the fries and the symbol for this is a match, accompanied by the slogan "fiery fries." Instead of picturing a match, the advertisers made the French fries themselves look like a match, with its end covered in ketchup.

In the next example (Figure 3.12) the featured product is photographed to look like the symbol, creating unequivocal visual resemblance between the two. The advertisement warns against the danger of drinking unpurified tap water; it uses the symbol "gun" as a symbol for danger. The visual resemblance between the gun and the shape of a water tap creates the fusion in which we feel that we are looking at a gun, even though we are looking at a tap. Message (danger), product (tap water), and symbol (gun) are seamlessly fused to one visual image that we "get" in one swift look at the advertisement.

Figure 3.12   Drink purified water. Caption reads "Hands up all those who still think tap water is safe to drink."

Brand advertising

The Metaphor tool is not limited to advertising products; it is equally effective in advertising brands. In almost every example we have presented up until now, a promise with relation to a specific product was advertised, but the name of the product remained on the sidelines. Many times, the name was an attachment to the advertisement in text or in the logo found in the corner of the ad. The logo itself can, however, play a more central part in the fusion, in the same relationship presented earlier with regard to the product.

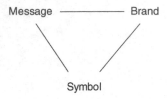

All three need to be fused in one visual image for maximum effectiveness.

As an example, the brand Macleans®, which is identified with toothpaste and with other products for dental and oral hygiene, wanted to deliver the message that their products whiten teeth and act against stains. The chosen symbol was an eraser and the fusion visually connected between the symbol and the name of the brand itself. We see Macleans® (brand) as an eraser (symbol and message) – in one visual image.

Look at the advertisement in Figure 3.13, which is for manufacturer of herbal teas. The company wishes to create a link between drinking its tea and calmness, delivering the message, or the promise, that tea products will induce calm. The symbol chosen to represent the value of "calmness"

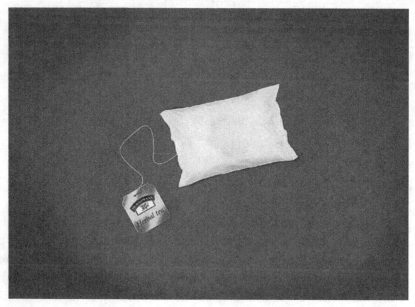

Figure 3.13   Pillow tea bag.

is a pillow. To create the fusion, the advertisers replaced the tea bag itself with a pillow – connecting the name of the brand, the category of products that carry its name and the effect of comfort, peace and rest in one visual image.

Brand recognition

One can only undertake a strategy of using the tool of Metaphor for brand advertising when the audience already recognizes the brand and knows the products the brand offers. This is a necessary precondition. If the audience is not familiar with the particular products of the brand and the advertisement is devoted to promoting the brand itself but does not provide information on the products that carry the brand name, the advertisement will clearly not be effective. Imagine, for example, the Macleans® ad described above, if we didn't know that the advertisement is talking about teeth cleaning products.

Therefore, fusion that takes place between the name of a brand, a promise and a symbol requires prior brand recognition and its immediate identification with a specific category of products.

In the example shown in Figure 3.14, the Lego® company advertises its products – which are identified with its company name. As this is a case in

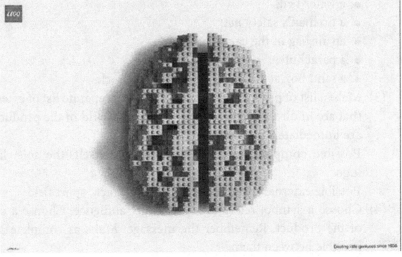

Figure 3.14   Lego® brain

which the category of a specific product is completely identified with the name of the brand, there is no need to mention the brand name specifically. Lego® toys immediately invoke the brand name. Thus, simply using the components of the toy in the ad creates the necessary fusion with the brand. The chosen message was that Lego® toys can be used as a tool for a child's cognitive development. The symbol chosen was a brain (Figure 3.14). The visual fusion is expressed through the pieces of the Lego® game forming the figure of the human brain. One visual image – brand (Lego® pieces), symbol and message (brain). Even though a specific product is not advertised, the message about the Lego® brand is clearly delivered.

INSTRUCTIONS FOR USE

(1) Precisely describe the message and the central idea in the message that will be connected to the product.

The advertisement is for Nike Air® professional sport shoes, with soles that contain air capsules.

The chosen message is shock resistance. Nike Air® is a shoe that acts as a shock absorber to the foot.

(2) Compose a list of symbols (objects, images, concepts) that are directly connected to the chosen message.

Symbols that represent shock absorbers could be:

- a spring or a coil;
- a trampoline;
- mattresses in a gym;
- a water bed;
- a fireman's safety net;
- an air bag in the car;
- a parachute;
- a sand box at the end of a long-jump track.

(3) Make a list of product components. Make a separate list of external things that are in direct contact with the Closed World of the product and thus are immediately connected to it.

Possible components could be: the shoe itself, the sole, laces, Nike logo.

Possible external stimulus: foot, running track, sport field.

(4) Choose a symbol relevant to the target audience. Choose a component of the product. Remember the message. Make as complete a fusion as possible between them.

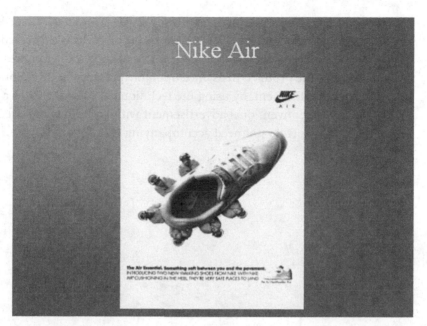

Figure 3.15   Nike Air sports shoes. Caption reads "The Air Essential. Something soft between you and the pavement."

- When the air bag near a car's steering wheel is inflated, it is in the shape of a shoe – or, a man runs with air cushions that are imprinted with the Nike logo.
- Children are jumping in a huge shoe.
- When a parachute opens, it is in the shape of a shoe.
- An athelete long jumps and lands in a shoe.
- Firemen hold a giant shoe instead of the safety net in order to catch the jumpers from above (Figure 3.15).

And some for the road

(1) The Metaphor tool is one of the most commonly used of all the tools in the creative department's toolbox. The strength of symbols and associations come from the simplicity and ease of application for every type of product, service or message to be marketed.

(2) The three-pronged process that creates a visual connection between the product, the symbol and the message was proven to be more creative than the common use of two of the three components. Usually, Metaphor fuses

product and message, as is expressed in informational advertisements. Sometimes Metaphor fuses message and symbol, as is expressed in more complex advertisements that have not been shown to enhance "memorability" of the product. Insistence upon the presence of three components in the advertisement, by using the technique of fusion, creates a memorable and unconventional advertisement with an immediate effect. Often, ads of this sort do not need accompanying text.

4 Subtraction

Minimalism wins creative results

One of the characteristics of a creative commercial, as was discovered by systematically analyzing advertisements that won creativity prizes, is Subtraction. Advertisements which favor visual simplicity over visual complexity and that use few words, rather than adopting a wordy approach, generally have a good chance of being considered more creative. The design of such advertisements is clean and focused; they deliver the message through reducing the background noise. Minimalist advertisements generally trade an attempt to dazzle by overuse of elements available to advertisers, with the use of wittiness or individuality. In the tool of Subtraction, minimalism is achieved by removing elements generally considered essential. Minimalism seems to work so well because we often notice objects only when they are missing.

Can we gain more attention for any advertisement by subtracting elements from it? The ads shown in Figure 4.1 are designed to raise awareness of the need to protect nature, offers a response to this question and will serve as an excellent introduction to the tool of Subtraction.

When subtraction is an addition

The Subtraction tool enjoins advertisers to eliminate one of the elements of an advertisement in order to achieve the effect of minimalism. Of course, Subtraction will not serve an advertisement if we omit information relevant to the client or make the identification of the product or brand difficult; the goal of the Subtraction tool is to provide added value in the cognitive, intellectual or emotional effect of the advertisement through subtraction

(a)

(b)

Figure 4.1   (a) An elephant never forgets; (b) Goldilocks and the three bears.

of essential elements of that advertisement; but how will subtracting something add to the advertisement?

Studying examples from the field of new product development will be the most straightforward way to approach this question. Sometimes, a radical step of eliminating an essential component from the design of the product or the service allows for the invention of surprising benefits or new market niches. Using this principle, eliminating one of the most essential elements of the television, its screen, caused the invention of a new product: a device from which we can listen to television, but can't see any programs. This product is actually manufactured and marketed in the United States and used by drivers who are interested in listening to their favorite talk show or soap opera while driving. The Israeli cellular telephone company, Pelephone, created a completely new product by eliminating one of the most vital elements of what one expects from a mobile phone service: they introduced a phone that did not have the ability to make calls. By taking this step, they began serving new and significant market segments. For example, parents bought this telephone for their kids, without worrying about an inflated bill.

Examples such as these and others depict the principle upon which Subtraction is based. When using the tool of Subtraction in new product development, a function or an essential component is subtracted to create a completely new benefit for the organization. With regard to advertising, the subtraction of an essential element must provide additional insight into the message delivered by the advertisement. The elimination of the essential element is not undertaken to harm the product, make it less valuable or to take out an element that is not wanted by clients in the market (for example – the principle of subtraction is not about coffee without caffeine or cheese without fat).

How Subtraction highlights the message

Look at the next example, a Chivas Regal® advertisement shown in Figure 4.2a. This advertisement makes generous use of the Subtraction tool and

(a)

THIS IS AN ADVERTISEMENT FOR CHIVAS REGAL

IF YOU NEED TO SEE THE BOTTLE,
YOU OBVIOUSLY DON'T MOVE IN THE RIGHT SOCIAL CIRCLES.

IF YOU NEED TO TASTE IT,
YOU JUST DON'T HAVE THE EXPERIENCE TO APPRECIATE IT.

IF YOU NEED TO KNOW WHAT IT COSTS,
TURN THE PAGE, YOUNG MAN.

(b)

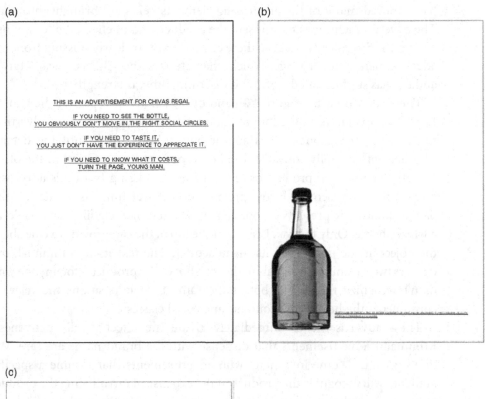

(c)

THE VIEW FROM THE TOP.

Figure 4.2   (a) Reputation; (b) No label; (c) View from the top.

has removed many of the advertising elements generally brought into play. The advertisement does not present the product, and in effect, it does not use any image. Specifically because these elements are so clearly missing from the advertisement, the message, which characterizes the Chivas Regal® target audience as sophisticated, elite, and discriminating, is strengthened.

The next two parts (Figure 4.2b and c), which are also part of the Chivas Regal® campaign, used the tool of Subtraction, as well. In the first advertisement below, the product does appear but without any identifying details. The element generally considered rather important if not critical in the other advertisements – the product or brand name, is missing from this advertisement. The advertisement below provides even fewer hints as to the identity of the product: it presents a top shot of a bottle, not readily identifiable as a whisky bottle. Only the wax insignia at the top of the cap provides a clue about the object in the image and its manufacturer. The text itself is minimal, but delivers two meanings: it signals to the top shot of the product – for anyone who didn't recognize the image, but at the same time it positions the relevant consumer at the heights of economic and social classes.

These advertisements immediately create the effect of distinctiveness. Minimalist advertisements that dispense with the brand name are rare. We don't generally come in contact with advertisements that assume a specific audience will recognize the product, at the expense of introducing the product to those who will not recognize it. However, even when taking the radical step of eliminating the name of the brand, an effect is created that directly empowers the advertisement's message: the Chivas company is positioning the product as appropriate for a closed clique – a product for those who have reached an elite status. Only members of this exclusive group will be able to appreciate the taste of this fine whisky – and only these same people are distinguished enough to recognize the product in the advertisement. It would be reasonable to assume that people who do not immediately recognize the brand in the advertisement will make it their business to find out what the brand is. Most probably, the company assumed that those who don't have anyone in their surrounding who could distinguish the brand couldn't afford their product anyway.

Subtraction can be used on any element – and partially

The Subtraction tool is not only meant for eliminating the product or brand name. This tool works with subtracting any of the essential components of

Figure 4.3   What 80% off actually looks like.

the advertisement: visual, logo, text, name of brand, picture of the product, or even the advertising platform. In the following example (see Figure 4.3), which advertises price reductions for Daffy's, an American discount clothing store, most of the sign itself on which the advertisement was placed, was bravely subtracted.

This example is especially interesting: first of all, it is an excellent example of the immediate effect created by use of the Subtraction tool. It grabs the attention of passers-by in a very creative way and at one and the same time, with the very same technique, it delivers the message of deep price reduc tions. In addition, this advertisement actually combines the thinking process used with two different tools: the Unification tool, which manipulates the medium itself and the Subtraction tool, in which it is the medium itself that was "subtracted." These two worked well together in order to emphasize the chosen message.

It is possible to partially subtract the component you are choosing. So, in the above advertisement, for example – the entire medium wasn't subtracted. An essential amount of the medium was subtracted. When considering the subtraction of an element, you can determine whether to make a partial or complete subtraction of an element by evaluating how the message will best be served.

INSTRUCTIONS FOR USE

(1) Precisely define the message or promise that you are choosing to deliver.

The chosen message is: Cindy Crawford will appear in nude photos in this month's issue of Playboy.

(2) Create a list of the essential components that make up most advertisements or make a list of the elements of an advertisement that are usually
required by the client.

The list may include the following elements:
- title;
- copy;
- image;
- picture of the product;
- product logo or brand;
- name of product or brand;
- the medium on which the advertisement is placed.

(3) Systematically, subtract each one of the elements on the list from the
virtual advertisement, and find the way that its subtraction could emphasize or demonstrate the message that you are trying to deliver.

For the chosen campaign, we can test out the following subtractions.
- Take off the title – the copy can be composed as follows: When we saw
Crawford's pictures we remained speechless.
- Take off the copy – the message is that there is no need for words
(or articles); the picture says it all.
- Remove the visuals – the text is naked like Crawford herself.
- Eliminate the image of the product itself – we couldn't stop taking
pictures of Crawford for the new issue and didn't leave space on the roll
of film to photograph the magazine itself.
- Take off the logo – the bunny is so excited by the pictures of Crawford
that he had to go calm himself down in his bed – or Cindy Crawford is a
bunny of a completely new kind.
- Subtract the name of the product – you of course know which is the
only forum that could give you this experience.
- Get rid of the element of the medium – 95% of the page is missing as is
95% of what Crawford usually wears.

(4) Adopt the concept you chose and implement it on the design of the
advertisement. You can adopt the concept through partial or full subtraction of that component of the advertisement. If we test, for example, the
subtraction of the component "image," we will then test if partial or full
subtraction will do better work for the product's message.

It was decided to develop the idea of subtracting the image. We formulated
the idea behind it as follows: the text is naked – meaning without an image
to support it, as Crawford herself is without clothing. The message we
decided upon focused on the element of Crawford's nakedness: the fact

CINDY CRAWFORD, WITH ALMOST NOTHING ON, IN THE SEPTEMBER ISSUE. ❡PLAYBOY

Figure 4.4   Cindy Crawford.

that clothing was missing from her body. This idea can be delivered through having Crawford herself absent from the advertisement. Or through eliminating essential elements from Crawford's picture. Therefore, most of the image of Crawford's body is subtracted, leaving only her recognized commercial symbol that is well known by all the viewers – her beauty mark (Figure 4.4).

Unintuitive thinking

Using the Subtraction tool usually prevents advertisers from getting stuck in the same old ideas. Thinking through use of this tool forces the advertisers to use what is called "the unintuitive path of thinking" in their cognitive thinking process and helps them avoid expected outcomes (Figure 4.5).

When we implement the creative process we make decisions, many of which are intuitive and not conscious: we choose paths of specific thought through inertia, habit or simply through complacency. These regular methods of thinking have been defined for us through positive experience with ways of thinking that proved themselves in the past.

The path of unintuitive creative thought, expressed in Figure 4.5, forces us to choose paths of thought that we usually avoid. A choice of this type enriches the mind as well as marketing thinking and serves as a significant tool for designing an advertising message that can surprise the customers and the competition.

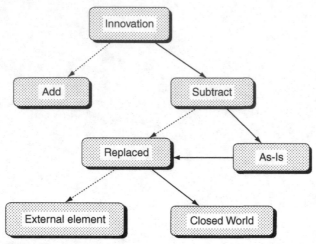

Figure 4.5   The unintuitive path of thinking.

The main junctions in the thinking process of Subtraction

(1) The first junction occurs when the advertiser needs to decide whether to subtract or to add. When we ask advertisers or customers to generate ideas in order to improve an ad, suggestions usually include adding elements, enriching components of the message – or adding additional messages. The "natural" tendency of thinking is to add information, to emphasize additional promises or to enrich the advertising message. Removing components is usually only undertaken to get a cleaner graphic design. However, elimination of components can serve as a source for renewed creative thinking about the advertisement. At the first junction of unintuitive thinking we will remove a central component of the advertisement or diminish its place in a way that will serve the chosen message.

The examples shown in Figure 4.6 are a part of an incredible campaign for Absolut Vodka®, a campaign that blurred the boundaries between creative advertising and a work of art. The entire campaign was based on creating manipulations by using the bottle itself and connecting it to a recognized cultural object. Many of the advertisements in the Absolut Vodka catalog are based on subtraction of generally accepted advertising components. In Figure 4.6, the image of the entire product is made to disappear as an illustration of the cultural symbol of Houdini; in another, the component of copy is removed as well as the title. In fact, the only thing that appears in the advertisement is a visual of the bottle itself, without addition.

(a)

ABSOLUT HOUDINI.

(b)

Figure 4.6 (a) Absolut Houdini; (b) Absolut Vodka

(2) At the second cognitive junction, after we subtracted an element in order to assist in the delivery of the message, we must decide whether it is better to replace the missing element with another advertising component or simply leave the component absent. Our normative tendency is to replace the element. We will be inclined to find another component in the system that can fill the task left orphaned by the missing component. However, we recommend that you don't automatically fill the empty space, but invest thought in what will serve the message best: leaving the advertisement without the element or finding a substitution for the missing element.

The following example is a part of the campaign of the Swedish furniture king, Ikea. Accompanying their end of season sale, the campaign delivered the message "At Ikea we cut prices." During the first roll-out of the campaign, the ads used copy to explain the price reductions. The images selected for the advertisement were items sold in Ikea that could be used for cutting, pruning, or splitting: an axe, an electric saw, tree clippers, etc (Figure 4.7a). During the second part of the campaign, shown in Figure 4.7b, copy was subtracted from the advertisement, but it wasn't replaced with anything. The advertisement simply showed the images of the physical products seen at the beginning of the campaign – the cutting

(a)   (b)

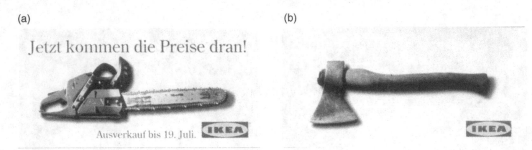

Figure 4.7   Ikea sale.

implements. The customers connected the visuals with the message that they had already seen. These images also created an immediate illustration of the entire message. The result was a very focused and effective campaign that carried a creative lesson: Subtraction can be used as a pay off for a completely different original campaign. A simple campaign of Unification, Metaphor or even a campaign that was not considered especially creative when it was first launched can be creatively turned into a campaign that uses the tool of Subtraction.

(3) If you decided not to replace components but to leave the advertisement with the missing item in order to amplify your message, you've completed the process. However, if you need to exchange a removed component for another component, you reach the third junction – choosing an element to take its place. At this point, your choice is between using an internal component that can be found in the Closed World of the product or service or an external component. An internal component will necessarily present the chosen message directly, while an external component is found outside the advertisement's or message's world of content. Thus, for example, if we choose to create an advertisement for sports shoes for professional athletes, using a visual component such as a running track, sport pants or Michael Jordan, we remain within the Closed World of the message. But if we include a famous model, a trendy night club or a tie in the advertisement, we will be stepping outside the Closed World. Intuitively, we may think that it will be more creative to step outside the Closed World of the product. However, it has been discovered that when we use symbols and elements from within the Closed World of the product or message, the results are generally considered more creative (this is discussed at length in the chapter on Unification). Thus, it is recommended to select substitutions from within the world of the product or message, although this too, may appear to be unintuitive.

Figure 4.8   Honey Nut Cheerios.

The example (Figure 4.8) advertises Honey Nut Cheerios Cereal®. It depicts "partial subtraction," briefly mentioned in the Cindy Crawford example (see Figure 4.4). In the advertisement below, it was decided to relinquish the copy – but not all of it. Because the English letter O appears in the name of the product and in its picture, it was possible to leave it out of the title. Notice that the missing letters weren't completely removed: they are present in the Closed World of the product and therefore they take the place of the partially missing element.

An additional example of an exchange that occurs within the borders of the Closed World can be seen in the advertisement shown in Figure 4.9 and sponsored by Amnesty International, the international human rights organization. Although the images were completely removed in this advertisement, they were replaced by short copy describing the horrors that would have been seen in the missing pictures. The original removal of the picture was to serve the message – just because you don't see something happening, doesn't mean it is not happening. However, when they came to the next junction of thinking – whether to leave the missing element absent or to substitute it – they decided that the advertisement would be much stronger if the text would describe what they weren't seeing. The exchange is accomplished within the Closed World of the components of the advertisement. They then added an additional element in the design: the "missing pictures" – or the pictures that were "banned" – are arranged within what looks like an undeveloped roll of film. This, too, was an exchange of the component

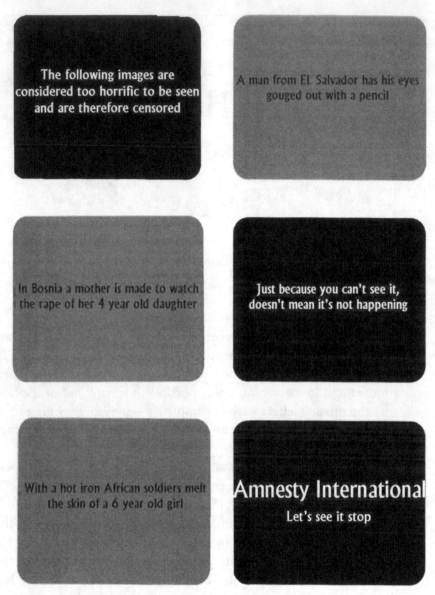

Figure 4.9   Just because you can't see it, doesn't mean it's not happening!

from the Closed World of the message. The film is viewed only in its potential; the pictures exist but we can't show them. This exchange depicts a combination of the use of the Subtraction tool and the Metaphor tool, because the image itself is subtracted, but in its stead we include a component that is a symbol for pictures, and this symbol serves and amplifies the given message.

(Be careful with the Kaminomoto)

Figure 4.10   Be careful with Kaminomoto.

The next example is an advertisement from the school of Neil French, the renowned copywriter responsible for the Chivas Regal® advertisements presented earlier (see Figure 4.2). The advertisements below promote a very difficult to advertise product: a hair growth product. A product of this sort, like other products and services that provide a solution to an unpleasant or discreet personal problem (such as weight loss products, products for improving sexual performance or breath freshening sprays) are not easy to advertise as potential customers do not like to be reminded of their need for a product of this type. We have discussed advertisements specifically designed to handle such situations at length, in the chapter discussing the tool of Activation. However, here is another alternative (Figure 4.10): if we show a bald man in an advertisement, many customers will choose not to identify with him and will prefer to ignore the advertisement. Advertisements for this type of product are most effective when gentle, humoristic, and implied. The image of the product, the headline and the specific visual mention of the problem solved by the product were all eliminated in the Kaminomoto campaign. The elimination of these elements corresponded with the message of lack of hair. As a message of this sort can be perceived as somewhat hurtful, French chose another bald object to portray in the advertisement, instead of a head.

And some for the road

(1) The Subtraction tool in advertising is based on breaking commonplace creative thinking patterns through subtracting an essential element in the advertisement, like the image, the copy or the name of the product, in a way that will serve the message delivered by the advertisement.

(2) Thinking through use of the Subtraction tool systematically evaluates each of the essential elements that appear in the regular advertisements and checks methodically how removing each one of them could possibly lead to emphasis of the chosen message.

(3) Subtracting elements from an advertisement in order to promote the effectivity of the advertisement and its creative value goes against the intuitive process of adding components in order to fulfill the same goals. But the research at the basis of this method proves that minimalistic advertisements that have undergone the process of Subtraction are actually recipients of greater creative appreciation. There are a few reasons for this data.

   • The missing element draws the viewers' attention and because of this, advertisements that have used the Subtraction tool elicit a greater attention span from viewers.

   • The component is eliminated with the express purpose of emphasizing and strengthening the message. Therefore, advertisements of this nature effectively correlate between form and content.

   • Many times advertisements that use the Subtraction tool elegantly overcome obstacles and effectively deliver images or messages that could be viewed as hurtful or unpleasant (like, for example, advertisements for products or services that solve a problem of which the customer is not interested in being aware: getting fat, balding, and others.)

   • Advertisements that used the Subtraction tool stand out in the sea of advertisements that are rich and abundant with detail and elements. It appears that less really can be more.

5 Extreme Consequence

Avoiding cliché – and still highlighting a product's promise

Advertising's biggest enemy is cliché – and there is no tactic in delivering messages as rich in cliché as the exaggeration of a product's promise. It doesn't matter whether the exaggeration is done in a banal, manipulative, careful or entertaining way. It seems that there is no way to avoid cliché if you want to talk about a product's promise.

Most groups of marketing students, company managers or non-advertising professionals, when asked to come up with an idea for an advertisement, will suggest highlighting how great the customer feels while using the product – or how well he's done because of its use. In short, they will recommend exaggerating the promise. In these ads, drinking a soft drink will transform the client into a young and adventurous person and putting money into a specific investment fund will completely obliterate any financial worries a bank client may have. This technique is intuitive to people seeking to convey a message about a product and is an extremely popular tactic in advertising because it clearly stresses the promise or benefit of a product, making the message of the ad very easily accessible. Unfortunately, it is also the least credible form of advertising. It is a cliché.

Our potential customer is not naïve. They've seen everything, and they've also bought products and services that promised heaven and earth and didn't necessarily deliver. Yet, it is possible to communicate the benefit or promise of a product without falling into the trap of sounding incredulous.

Consequence replaces promise

What we are suggesting is as follows. To replace the common cliché of extreme promise with another, similar extreme, but one that will solve the problem

Figure 5.1   Sonicare toothbrush for sparkling teeth.

of credibility. The name of the tool is Extreme Consequence. What we see in ads characterized by this tool is the absurd result of using a product or having it.

Here's an example of how Extreme Consequence differs from exaggerating a product's promise. There have been innumerable advertisements for dental products that promise white teeth. These promises are usually of two types: (1) the "scientific" message, which presents data on how quickly teeth are whitened by using this product or how many percentage points more white they will be when the customer regularly uses the product; and (2) the emotional message, which shows how exquisitely beautiful teeth become after using the product – usually displaying them on the lovely face of a wide-smiled model. Look at the advertisement in Figure 5.1, however. It is based on the same promise – white teeth. Yet, it makes use of the promise in a different way

In this ad, the benefit of the product, or the product's promise, is immediately understood. So what distinguishes between this and the many other advertisements that exaggerate a promise? What turns the above advertisement, as well as the one in Figure 5.2 for Aquasoft® pants, into creative award winners?

The answer is simple, but in the processes of creative thinking and creative ideation, it is not always intuitive.

Instead of exhibiting the common schema:

So comfortable, you might forget you have them on. AQUASOFT

Figure 5.2   So comfortable, you might forget you have them on.

Core positive attribute – positive benefit – positive result of enjoying or realizing the benefit
Extreme Consequence ads take another route:
Core positive attribute – positive benefit – *negative or unexpected consequence of enjoying or realizing the benefit*

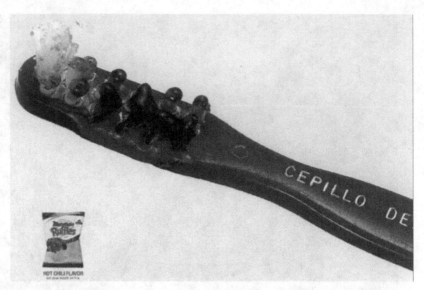

Figure 5.3   Burning toothbrush.

Creative ads that subscribe to the logic of Extreme Consequence do not follow the familiar path of exaggerating the product benefit. Instead, what they do is present an unexpected, accidental, and often negative scenario, arising as a consequence of the product's positive attribute (see Figure 5.3). The element of exaggeration is essential to the effectiveness of the tool. The scenario in the ad must be unequivocally absurd, unrealistic and over-the-top. If not, it risks shooting itself in the foot. Not by showing some unbelievable consequence of using the product, but rather by raising consumers' fears about the implications of using the product.

The Dunkin' Donuts® commercial drives home the message that even in the busiest of days, there is always time for a breakfast bagel and a cup of coffee (Figure 5.4). The commercial depicts a car chase shot in the aerial style of America's Most Wanted. We see the suspect pulling up in front of a Dunkin Donuts® store while still in full pursuit. He jumps out of his car and rushes into the store, pursued on foot by the police officer. We then see both emerging with bagel and coffee in hand only to jump back into their respective vehicles and the pursuit continues.

The ads in Figure 5.5 are taken from a brand campaign for GT®, a well-known US cycle brand. The brand's message is "speed." Everything connected to GT® is affected by speed. Speed is the value upon which the company is run, the chief value of its product. However, instead of dramatizing this through

Figure 5.4   Police chase.

(a)   (b)

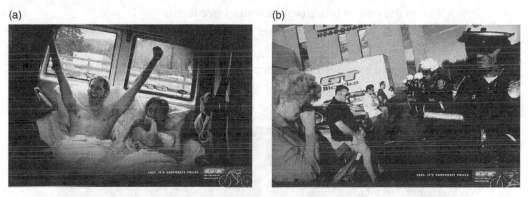

Figure 5.5   Fast: (a) sex; (b) speeding tickets. Caption reads "Fast. It's corporate policy."

absurd scenes of bikes overtaking fast cars on the open road, customer service representatives answering calls before the phone rings or staff with superhuman reaction times, the ads show how the company's policy of speed can also get in the way and cause problems.

By using Extreme Consequence in such a way, GT® does not expose itself to promises that it cannot substantiate nor expectations that it cannot fulfill. Extreme Consequence helps create an ironic and tongue-in-cheek tone of voice. And yet, importantly, this humor, which is so central to the campaign's appeal, does not come at the expense of diluting the brand's core characteristic: speed.

INSTRUCTIONS FOR USE

(1) Choose an outstanding positive attribute or characteristic of the brand, product or service.

The chosen product is canned tomatoes.

The core product attribute is the fact that the product contains just tomatoes and is free of preservatives, which gives it the taste of fresh, natural tomatoes.

(2) Formulate the benefits coming from the positive attributes of the product.

You don't need to use fresh tomatoes.

(3) Think of scenarios where the benefit (mentioned above) leads to a negative, problematic or unexpected result. Try and make sure the situation is really absurd.

- A house-wife tries, unsuccessfully, to rinse a can of tomatoes while preparing a green salad.
- A skinny model munches on some fresh vegetables but when she gets to the can of tomatoes she cracks her teeth.
- Cubes of canned tomatoes are found in the pasta sauce instead of tomato slices.
- Cans of tomatoes are thrown at a stage comedian who fails to amuse the audience (Figure 5.6).

Figure 5.6   Real tomatoes!

Figure 5.7   Venus fly trap.

When applying Extreme Consequence it is preferable to come up with negative exaggerated consequences. However, sometimes a surprising or unexpected consequence instead of a "negative" one is as effective (Figure 5.7).

Any ad for push-up bras must promise a fuller, more prominent bust. You can imagine many possible ways of expressing this promise, ranging from the too obvious to the too vulgar. The copywriting team who created the ads were asked to steer clear of hackneyed solutions or sexual clichés. What they came up with were a series of everyday scenarios in which the push-up effect of the bra results in totally unexpected consequences for the wearer. An example is presented in Figure 5.8.

These examples lead us to a further use of Extreme Consequence. It is one that we have already touched upon earlier in our discussion, but with one important difference. Instead of focusing on the consequence that comes from enjoying the core product benefit, the ads lock on to what is actually a relatively insignificant feature of the product, and then explore the implications of this. The next example presents the talents of an advertising photographer. It neither directly emphasizes his grasp of the marketing genre nor his professional acumen nor his experience. What it does emphasize, however, is the requisite know-how of any advertising photographer to set up a photograph within the constraints of magazine printing. The surprise comes not from his ability to do this, but from negative implications of

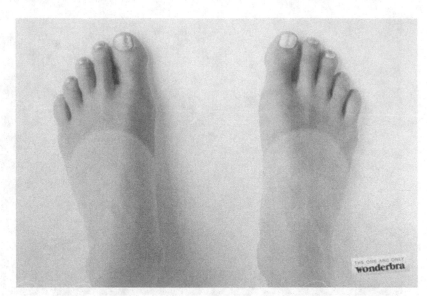

Figure 5.8   Feet in shadow.

Figure 5.9   School photograph.

this approach when applied to more conventional photo assignments such as that shown in Figure 5.9. In any case, the take-out of the ad, is that Julian Wolkenstein, is an advertising *specialist*.

An ad for Lexus® that again fits into our atypical pattern.

The ad shows an egg as part of a breakfast meal, yet instead of depicting its nutritional value, it is "examined" as though it was a car – i.e. one yolk = room for only one passenger; egg whites and yolk = only comes in two colors; hard shell = no ventilation.

The banner on the ad excuses this thorough examination by saying "Our Engineers Can't Even Enjoy Their Breakfast."

Instead of talking about the car's safety, its design or driving performance, the focus of the ad is on a more minor aspect of the brand. The engineers are so obsessed with their work on the car that they cannot separate it from the rest of their life.

In choosing to focus on this particular minor quality, the advertisement actually stresses the product's central qualities – if the cars are made by engineers of this caliber, they are necessarily safe, excellently engineered, and provide a smooth ride – although the advertisement doesn't promise any of that. The viewer makes the connections in his mind when reading the ad. Additionally, the absurd situation actually shows a positive, if certainly unexpected, result of the minor trait. However, it doesn't in any way make us feel that the product is touting its promise or tooting its own horn in an unreliable way.

Thinking process for presenting a minor trait

Again, the intuitive process in presenting the promise of a product is *Core positive attribute > Positive benefit > Positive result of enjoying or realizing the benefit*

When creating an Extreme Consequence of a minor trait, we replace it by *Minor positive attribute > Positive benefit > Unexpected situation (positive or negative) of enjoying or realizing the benefit*

Provide a new approach to a product

In the following award-winning television commercial for Miller Lite® beer, the advertisement focuses specifically on the way you open the bottle (Figure 5.10). The advertisement shows a heavyset man looking at the closed bottle of beer, with a worried expression on his face. Suddenly, he breaks out into a clumsy rendition of the Twist. The bottle of Miller Lite® beer is constantly on screen, a focused element in the commercial. The man

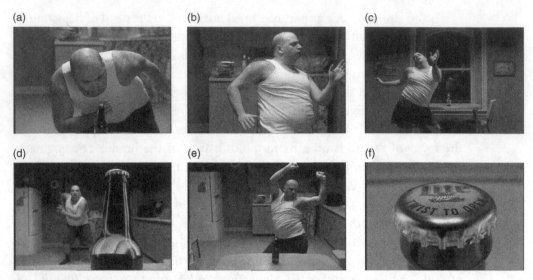

Figure 5.10   Twist to open.

continues his funny dance for a few moments when he looks expectantly at the bottle of beer. We then see the inscription on the bottle cap – "Twist to open."

The advantage of focusing on a minor attribute is that it allows the advertiser to approach a product in a completely new way, while also reminding the customer of the product's hidden advantages and its added value. Viewers, who already identify the central value with the product and immediately recognize that central value, will now be able to add this hidden value to their perception of the product.

So, for example, Miller Lite® viewers, who already know that it is great, light beer because the product is recognized in the market, now also know that it has a twist top. More importantly, they stopped and perceived the product in a new way. They had to give the product attention. In the mature stage of a product, when advertisers are interested in creating a new perception of the product or the brand and reinvigorating its image in the consciousness of the consumer, focusing on a "new" attribute in the media, even if it is minor, can do the trick.

Focusing on a minor attribute is appropriate when promoting products that lead and define their categories. The category itself must be identified with the product. A minor attribute should be stressed only when the core attribute – and the benefits derived from it – is very well known and recognized by the audience.

Many times, Fusion, the act of physically fusing the brand into the advertisement, an act which is required for brand recognition – is too complicated to implement in advertisements that are based on a minor attribute. Even though consumers remember the advertisement, they don't always recognize the specific brand advertised because the connection between the minor attribute and the product is not clear enough as when compared to the core attribute of the product. Therefore, when selecting the Extreme Consequence tool on a minor product attribute, there is special importance in stressing Fusion.

This can be done in a number of ways. The brand can be incorporated as a central part of the composition of the advertising image and not as a secondary component; Fusion can be created by using an object within the advertisement itself according to the principle of metaphor, whereby an audible or visual connection between the name of the brand and the central elements of the advertisement can be created.

INSTRUCTIONS FOR USE OF EXTREME CONSEQUENCE, USING A MINOR TRAIT

(1) Define a minor product attribute which offers a positive benefit.
The product is a Volkswagen car.
One of its minor attributes: the strength of the built-in air conditioning. It would be reasonable to contend that no potential client will make a purchase decision due to this.

Original Volkswagen air-conditioning.

Figure 5.11   Magnets on a car.

(2) Think of absurd situations, in which the minor benefit of the product could be central. The situations can have a positive or negative result.

To arrive at such situations, we ask ourselves: What can be the result of over-air-conditioning a car?

- An Eskimo family lives in it.
- A zoo houses a polar bear or an artic penguin inside.
- On a hot day, family members wear coats, gloves and scarves when sitting in the car.
- The car is filled with ice cream and other frozen products.
- The car replaces the family refrigerator – and they place their magnets upon it (Figure 5.11).

And some for the road

(1) The best way to avoid cliché is self-effacing irony. Use it when you present a situation in which the benefit that results from the promise of the product brings a negative result.

(2) Traditional exaggeration of the promise does not engender trust; when the message you choose reflects this understanding, it is perceived as much more credible.

(3) It sometimes makes sense to focus on a minor quality or product attribute – or one that is not intrinsic to the product – and exaggerating the unexpected result of this. Doing this may help shed new light on the product and offer a fresh way of talking about it.

6   Absurd Alternative

Focus on the benefit – not the product

Some of the most creative ads illustrating product results or solutions are those that do not focus on the actual product. Leaving the product out of the center of an advertisement presents a conceptual and design opportunity, allowing the advertiser to extol the benefits of a product without falling into a non-credible exaggeration of its qualities.

The product is also absent in ads that use the Inversion tool (see chapter 7). In these kinds of ad, the fact that the product doesn't exist in the absurd situation of the ad, makes it crystal clear to the viewer how important the product is and how much it is needed. In a similar vein, when advertisements created with the help of the tool we are presently exploring, Absurd Alternative, do not feature the product, they are free to focus on the benefit. Ads that use the tool of Absurd Alternative offer the viewer additional possibilities from within the Closed World of the product to achieve the same benefit offered by the product, but with a twist.

Advertisements using Absurd Alternative engage in the following dialogue with the client: "Maybe you *don't* need the product in order to get the benefit that it promises. There are other ways to get the same benefit – and it is our intention to present these other ways to you now." But, these alternative paths to achieve the benefits, suggested by the product manufacturer with extreme generosity, are, of course, outlandish and silly.

Generous irony in absurd situations

The following press ad struggles with one of the hardest possible products to advertise, that will be briefly touched upon in the discussion on the tool of

Figure 6.1   Life insurance promotion.

Inversion – life insurance. Considering absurd alternatives to purchasing life insurance requires exploring other possibilities that will provide the same benefit as life insurance. The possibilities are endless. The whole family can be practical and work together by robbing a bank when they notice that the time of one of the older family members is approaching. The deceased can add a clause to his will requiring the mourners coming to his funeral to purchase an entry ticket. Family members can find a double who will work in place of the deceased and continue to earn a salary. The advertisement in Figure 6.1 offers yet another possibility.

This ad was the recipient of creativity awards because it avoided the vulgarity of visually presenting the difficult condition of the family members of an individual who did not invest in life insurance. The advertiser could have chosen alternate situations. It could have shown a family's hardship in the absence of life insurance, mentioning how family members will have to work hard for a living or that they will have to marry for convenience to the wrong kind of people. However, these types of situations raise aversion and distrust in the client. One of the primary reasons for this is that the situation is not sufficiently exaggerated. This distrust is side-stepped in the humorous advertisement in Figure 6.1, although the message remains identical.

To reach this goal using the Absurd Alternative tool, the situation chosen must be patently ridiculous. The more insane the situation, the clearer

Figure 6.2   Cotton wool or a safety conscious car.

it becomes to the viewer that there really is no realistic or feasible substitute for the product being advertised. The advertisements in Figures 6.2 and 6.3 offer additional examples of this.

Another example of a creative use of Absurd Alternative can be found in the following campaign for hair growth products Figure 6.4. The commercial does not expound on the wonders of the product. Instead, it offers another option for getting hair. The man in the commercial surreptitiously shifts a clump of his airplane seatmate's hair onto his own head. He quickly finds a mirror and proudly takes a long look at his own head, decorated with his neighbor's mat of hair. This is an absurd, funny and unexpected – albeit only temporary – way to acquire the benefit inherent in the product's promise. It is the ridiculousness of the improvised "solution" that makes it clear that there is no real alternative to buying the product. And the beauty is, of course, that the advertiser need not say any of this. It is understood on its own.

Thinking processes for using the tool

In order to explain how this tool can be practically implemented, let's look at an example. A large telephone company wishes to launch a campaign boasting cheaper international phone calls. In order to use the Absurd Alternative tool, the advertisers will need to begin by creating a list of

Figure 6.3   The more humane way to keep him fit.

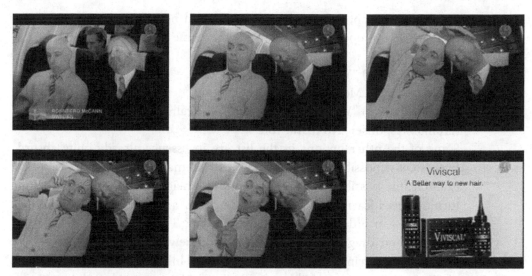

Figure 6.4   A better way to new hair.

realistic and unrealistic alternative routes to get cheap or free international communication.

- Use pigeon post.
- Send a message in a bottle.
- Swim across an ocean to get the message through.

These are amusing and absurd possibilities but none of these alternatives come from the Closed World of the advertisement's message. This "Closed World" can be explained as anything within the context of the advertisement's message. In this case, it would be anything that has to do with cheap international telephone calls. If we limit our thinking only to possibilities of free international telephone conversations we can compose the following possibilities.

- Call collect.
- Call in secret from a neighbor's house or from your place of work.
- Break into a telephone switchboard.
- Put a coin with a string into a phone booth that takes coins.
- Talk in agreed code to save time.

The possibilities are numerous, but before we can choose any of these as our option for delivering our message, we must be sure that our options are sufficiently silly. Any possibility that is not completely outrageous must be exaggerated. A collect call, for example, is an entirely reasonable way to talk inexpensively abroad. Therefore, a situation that would use this possibility must be exaggerated. For example, perhaps you can suggest a situation in which the relevant information is transmitted between one caller and the other while the operator is asking for the name of the person making the collect call.

(Recorded voice) "You have a collect call from…"

(Human voice) "Mom, everything is OK"

(Recorded voice) "Would you like to take the call?"

(Human response) "No thank you."

Some of the other options in the list are in essence already quite absurd. Even so, it is possible to give them an extra dimension of humor. Take, for example, the possibility of secretly using the neighbor's phone. How about turning this idea into a script that describes a couple's joint mission of seduction – in order to get to that telephone? The woman can seduce the neighbor and drag him into the bedroom; viewers will then discover her husband slipping into the house, in order to call his mother. Ridiculous scenes of this type are very accessible once the basic idea for an Absurd Alternative is reached.

Telia®, the Danish telephone company, ran an award-winning TV campaign in which each commercial presented a different inexpensive way to call internationally using a telephone. In one commercial, the man makes his call speaking at lightning speed without pausing for breath then hangs up immediately upon saying the last word. In another, the caller makes weird noises throughout the conversation, imitating problems on the line. He whistles, squalks, and squeaks, before asking the person at the other end if he would mind calling back because of the awful line.

For example, the ads in Figure 6.5 use Absurd Alternative in order to communicate one of the core benefits of shopping at the furniture retailer, Ikea.

(a)

(b)

Figure 6.5   (a) Toilet seat frame; (b) A sofa made of used tyres. Caption reads "It's true you can have it cheaper."

Additional uses for this tool

The Absurd Alternative tool can also be used as a device to assess the effectiveness of a message. Any message that cannot be worked on by the Absurd Alternative tool is not single-minded enough. Or, possibly, the benefit that the message describes is not special enough! If a benefit is not defined specifically enough, it will not be possible to discover specific alternatives to get it. For example, if an advertiser chooses a message of the type "it's the best possible product" or "everyone wants our product," it will be impossible for him to find alternative possibilities to meet the same goal. The message is so general and unfocused that there really is no other way to achieve the benefit set out in the message. The problem inherent in such a promise will be quickly exposed by working a message through the Absurd Alternative tool.

When leading workshops, we insist on running every message through the Absurd Alternative tool. This way we make sure that it stands the test of "being single-minded" before we begin the process of creative ideation. If the message does not stand up to the test, we view it as a warning sign. The message needs to be reformulated in a more focused way. The tool can also help choose between a number of message options.

We held a workshop for an international company that cultivates and distributes flowers to flower markets all over the world. Our goal was to brand a specific type of flower. The messages suggested by the planners included possibilities such as flexibility (the flower's stalk is flexible; the flower is adaptable to every bouquet), nobility, (the flower will make any bouquet more refined), and user-friendliness (the flower is long-lasting and easy to care for, with little upkeep).

In order to decide between the various messages, we tried to plan an advertisement using Absurd Alternative. Within the Closed World of "fresh flowers," we found alternatives for nobility almost immediately. We can achieve "nobility" by decorating the heads of simple flowers with symbols of nobility or with a royal crown. Or we can illuminate the flower with a crystal chandelier. For "flexibility" we found fewer alternatives (put a flower on a rubber band) and for "user-friendliness" it was harder to find alternatives within the Closed World of fresh flowers that would provide the same benefit. Beginning our creative task by using this tool in a way not traditionally considered creative, assisted decision-makers in choosing between alternative messages. This tool also allowed us to define the message and the benefit that the product offers with greater precision.

INSTRUCTIONS FOR USE

(1) Compose the message of the product or the service in the context of a benefit (e.g. if you buy the product, this or that will happen.)

The product is dandruff shampoo.

The benefit: using the shampoo will eradicate the problem of seeing dandruff on the head and body.

(2) Think of alternatives within the closed world of the product that will provide the same benefits that the product promises. When imagining alternatives, make sure that the possible benefits are completely achieved. List the other possibilities that will ensure that dandruff will not be gathered on the scalp or be seen on the body:

- wear only white (Figure 6.6a);
- stand on your head all the time;
- wear a hat;
- shave your head;
- vacuum the dandruff from your scalp every morning (Figure 6.6b);
- always stand at a distance from other people;
- shake your head every couple of minutes;
- Color your hair white.

(3) Review the possibilities and exaggerate any alternative that is not absurd enough.

With regard to the alternatives that we raised, we can exaggerate alternatives such as shaving the head. In order to exaggerate this alternative, which in effect could be a reasonable solution to the problem, we can place the head shaving in an absurd context. For example, a man who suffers from dandruff joins a Tibetan monastery in order that he may shave his head in peace; a man is forced to live in the company of fascist skinheads in order to explain his shaven head – an act he actually undertook due to dandruff; a man joins the marines just to get rid of his dandruff problem, etc.

When the product should be in the ad

In the example presented in Figure 6.6, the product is positioned in a full half-page of the advertisement, right next to the alternative. This is not an ideal situation with regard to the viewer's visual attention. However,

(a)

(b)

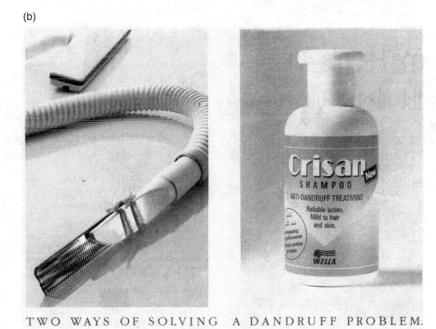

Figure 6.6   (a) A white wardrobe; (b) A vacuum cleaner.

sometimes, when the brand does not lead the category, the graphic tactic of placing two alternatives next to each other is the lesser of two evils.

As in the other tools that rely on exaggeration, the Absurd Alternative tool is best used on products that are clearly identified with their category or that lead the category. This is because the product has very little actual exposure in the advertisement, either visually or in a description of its attributes. If a product is not a category leader or identified with the particular category, it may be difficult to identify the particular product with the advertisement.

When using this tool, it is more difficult to achieve fusion with the product or brand. Again, this is because in most instances the product or service doesn't appear in the advertisement at all. In other cases, its presence is minimal. The viewer will remember the idea – but not necessarily the brand. Therefore, the tool is appropriate for market leaders and for public information messages – social, environmental or political – that do not require any kind of fusion. Figure 6.7 presents alternative ways of highlighting an important environmental and health issue.

(a)

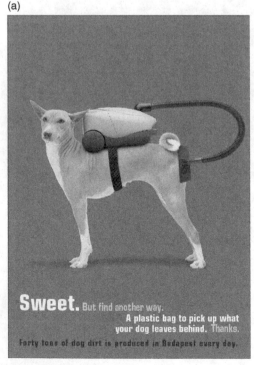

(b)

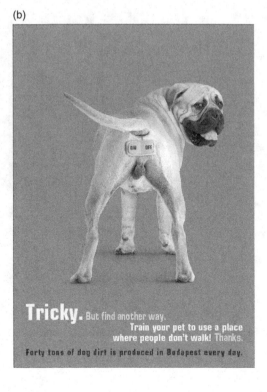

Figure 6.7   Clean Streets – ways and means!

And some for the road

(1) In order to assess if a message is focused enough, run the message through the Absurd Alternative tool and check if there are other alternatives for achieving the same benefit. If there are no other possibilities for achieving the same benefit of the message you chose, it is possible that the message is too general or not single-minded enough.

(2) An Absurd Alternative ad "generously" suggests to the consumer ways to avoid acquiring the product and still get the desired benefit. However, care must be taken to ensure that the alternatives are completely absurd. If an alternative appears even slightly feasible, try and make it more unreaslistic.

(3) Sometimes, the similarity between the Absurd Alternative tool and that of Inversion can confuse. The difference is simple. With Inversion, the benefit *can not* be achieved, whereas with Absurd Alternative the benefit is obtainable, albeit in a way that is clearly undesirable.

(4) As the product is not central to an advertisement using the Absurd Alternative tool, advertisers must ensure a strong product or brand fusion (unless the product they are promoting is a category leader or the advertisement carries a public information message).

7   Inversion

In marketing – there's nothing to fear but fear itself

In this next tool we will explore another technique of extremes that raises the question: what would happen if you don't have the product?

Many advertisements that attempt to answer this question will respond with: life will be bad, bitter, and lonely. These advertisements maintain that if you don't consume the product or service advertised, you will easily turn into a lonely, miserable, unaesthetic or desperate individual. But fear tactics, which were once quite popular in delivering messages, lose their effectivity as the consumer becomes more practiced in absorbing marketing messages and savvy in understanding economic motives. When facing an aware consumer, fear is not necessarily going to do the marketing trick. However, the Inversion tool offers a few principles for emphasizing the benefits of the product, as well as remarking upon the drawbacks of life without the product, that can serve as a useful alternative to the fear factor.

Let's start with an example. One of the products that is hardest to promote is life insurance. This type of service is problematic to advertise because it touches on a raw nerve: no one wants to think about death, especially not his own. With this in mind, when we ask advertisers to describe a situation that would be caused by not purchasing life insurance we usually arrive at the following examples.

- The family members come to the funeral dressed in rags.
- The family members are angry at the deceased at the memorial services.
- The family members are forced to work at hard physical labor to make a living.

These possibilities come from the associative paradigm of fear – if you don't buy the product, trouble will occur for you and to your family. However, in the present marketing climate, these ideas – in this rough state – would be considered vulgar and raise antagonism in the viewer. None of these scenarios would make a consumer feel positive or excited about the possibility of purchasing the product. Instead, they warn of a horrible future, a message that most would prefer to ignore.

The Inversion alternative

How, then, can we express the drawbacks of not having the product? How can we highlight the problems that will be caused by not having this product without threatening our viewers – and by so doing – stepping onto the minefield of lack of credibility?

The advertisement in Figure 7.1 solves this dilemma by completely exaggerating the vulgarity of the situation.

This advertisement invites a smile instead of resistance, specifically *because* it is not credible. We are surprised by the complete lack of congruity between our expectations of what should be written on a flower wreath and what is actually written. Similarly, we are amused by the advertisement's version of the results of not buying life insurance since they have been exaggerated to the extreme. Both those elements together make this an interesting advertisement. The advertisement jokes with the viewer. It doesn't threaten.

Figure 7.1   Why no life insurance. Caption reads "Don't worry, even if you leave them with nothing they'll remember you."

Figure 7.2   How to improve sight.

This extreme, non-realistic exaggeration can also characterize the message of the advertisement in Figure 7.2.

The message is, of course, that if you do not go to the optometrists advertised in the advertisement, your vision will deteriorate; but the situation chosen to represent the deterioration of sight is not realistic, the advertisement doesn't present an image that would seem to actually jeopardize a person's quality of life. As a result, the advertisement avoids cliché and most importantly, avoids threatening the client.

When using the Inversion tool in an advertisement, the advertiser reveals a situation that will happen to a viewer if he does not use the product; however he avoids frightening the viewer by employing situations that are completely unrealistic and will never happen or by suggesting absurd and funny ways to deal with not having the product.

Absurd situations make it real

Paradoxically, it is specifically because of the lack of realism in the scene that viewers consider the message more realistic (see Figure 7.3). This is not to

Figure 7.3   Auto-leveling. To the rescue.

say that they believe that the specific event will happen, but they are more inclined to believe in or to assimilate the advertisement's message. As the advertisement doesn't play in the court of negativity and doesn't try to shock with pessimistic results, it avoids the viewer's response of "it won't happen to me," which is a common response to advertisements that threaten or preach. Consumers tend to have reservations regarding advertisements of this sort. The scene described in this advertisement will definitely never happen, and because of this, it allows viewers to open their minds to scenes that possibly could happen.

Using Inversion on our old standbys

Inversion provides a decided advantage for campaigns that promote basic products. At its essence, the Inversion tool describes the absurd and generally funny condition of a consumer who does not have the product, when he is in a situation in which the product would be needed, or in a situation that would be created by not using the product. This is an angle that a consumer will not naturally consider. As such, the Inversion tool actually enables advertisers to present an old product in a new light. It is thus especially useful in a marketing strategy designed to remind consumers of their need for a particular basic product and how central it is to their lives.

Consumers take products they know for granted, as John Steel, the legendary planner from the Got Milk? campaign taught. He explained that basic products – milk, water, electricity, city infrastructures, etc. are not given attention by the consumer. Advertisers need to get the public to reconsider these products – to pay attention to them in order to relearn their value. It is not effective to repeat – again and again – that milk has calcium and is important for good health. The audience got those messages. Telling them again is not going to make them appreciate milk any more. For products like these, it is a more effective technique to work on the client's perception of these products. We want to make them look again at the product and re-evaluate its importance. Joni Mitchell sings what we all know: "You don't know what you got till its gone." It seems that situations that illustrate the absence of basic products – our old standbys that we think will always be there – may be more effective in promoting these products than simply repeating benefits that are already acknowledged and familiar to the audience (see Figure 7.4).

(a)

(b)

Time to switch to Toshiba professional lighting   Time to switch to Toshiba professional lighting

Figure 7.4   Highlighting: (a) Shell gas station and (b) Angus steakhouse.

Thinking process

With regard to the thinking process needed to develop the advertisement, Inversion is a mirror view or opposite view of "Extreme Consequence." We heighten the result of the non-use of the product to the absurd. Some of the principles that characterize the Extreme Consequence tool are true for the Inversion tool, as well. Two of them are now reviewed.

Exaggerate, exaggerate, exaggerate

It seems that the relationship between exaggeration and credibility is as follows. The more extreme, absurd, and non-realistic the situation you present, the more positively and effectively your message about the price paid for not using the product will be taken. The message in the advertisement shown in Figure 7.5 is about availability of new films. The absence of the product (a cable television channel) does not make the client bored, nor does it make him outdated. It will make films and their stars look like this to them.

The advertisement for deodorant presented in Figure 7.6 is staged in a crowded ballpark stadium. While doing "the wave," a man is left in his section entirely by himself so as not to offend the other spectators by the failure of his deodorant. The message of lack of deodorant has potential to come across as vulgar or offensive. Yet, with the use of exaggeration, the message is clear, witty, and inoffensive.

Another example of a commercial that expresses the severity of life in the absence of the product introduces us to a bashful young couple sitting side by side on a bench. They look at each other with desire and finally, they kiss. During the passionate and hot French kiss, the young boy's body begins to shake and to twitch. When the kiss ends, the boy is left on the ground, lifeless. The product is Vicks® – which opens a stuffed nose.

In the example presented in Figure 7.7, not using the product won't cause a truly problematic, traumatic or life-threatening situation. It can just be annoying. Therefore, the message of an advertisement of this type clearly won't be vulgar. The use of Inversion in an advertisement of this type, therefore, can just play with the message – make the message stronger and funnier by presenting it in an even more absurd way. This is what Virgin Air did in the advertisement for their newly, developed in-flight entertainment program.

Figure 7.5   Now films won't have to wait years to get on your TV.

Dealing with a minor trait

Sometimes, situations devised to show the negative results of what occurs in the absence of a product or service can lead to bragging. In order to avoid this pitfall, choose a relatively minor benefit of the product on

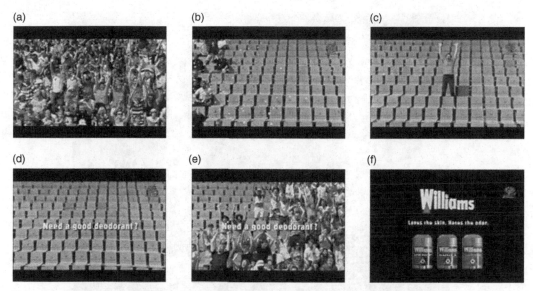

Figure 7.6   The wave. Caption reads "Need a good deodorant? Williams: Loves the skin. Hates the odor."

which to concentrate. A minor trait would be one that could never serve as the product's central focus of attraction by itself; it is not essential to the product. Then, using the Inversion tool, show how lacking this minor trait can be a real problem – thereby showing that the minor trait is actually very essential.

The advertisement in Figure 7.8 for an institute that teaches Spanish does not concentrate on the central benefit: cultural adaptation, knowledge, advantage when traveling to Spanish speaking countries – but instead chooses to concentrate on how not learning the language could affect your social life – a minor trait … .

Thus, for example, if the product for which you are creating the advertisement is a gym, you have an alternative to concentrating on building possible situations that draw attention to the central benefits of the product, such as "slimming," "body sculpting" or "makes you feel healthy." Inversion in this sense can possibly plunge into unpleasant banality in the form of – if you don't use the service, you will be unattractive or you won't feel good. However, gyms have many other benefits that are not necessarily fundamental but can illuminate the product in a new light. It can express a benefit not usually connected with a gym. In this way, it adds benefits that the public wouldn't have necessarily considered when thinking of a gym.

Figure 7.7   In-flight entertainment: "Stories About My Grandchildren."

We can consider benefits such as:

- in a gym you can watch TV without being bothered;
- in a gym you can meet people;
- in a gym you can fall in love or meet flings;
- in a gym you can take a rest from the family and the kids.

Using the Inversion tool on any of these benefits is not at all threatening and will not be vulgar. In particular, no situations based on such inversions

Figure 7.8   Why you should learn Spanish.

can appear aggressive. Similarly, the viewers already know about the primary benefit. It would be reasonable to assume that an attempt to directly emphasize the central attraction of a gym will most probably fall into the category of things the audience has already seen. However, an advertisement that features a bored woman, forced to see the football game with her husband and his noisy friends; or instead, a raucous scene at home, when the kids, neighbors, and family don't allow the father coming home from a hard day's work a bit of precious rest; or, perhaps – an introduction to a man, stuck in his apartment, who is engaged in a deep relationship with his goldfish – any of these scenes allow us to catch two birds with one stone: while emphasizing the secondary benefit, they somehow make the primary benefit, with which the audience was already familiar, all the more clear to the viewer.

In the advertisement for a headache pain pill shown in Figure 7.9, the advertisement's promise works in a similar vein. It does not threaten that anyone who doesn't take this pill will suffer from terrible migraines – at least this is not the threat that we see in the advertisement. The result of not taking the product would be that you have a great excuse for not having sex. This advertisement provides an excellent example for what this tool accomplishes. People know that headache pills help against pain. Precious little can be added to us on that score – we've heard the message

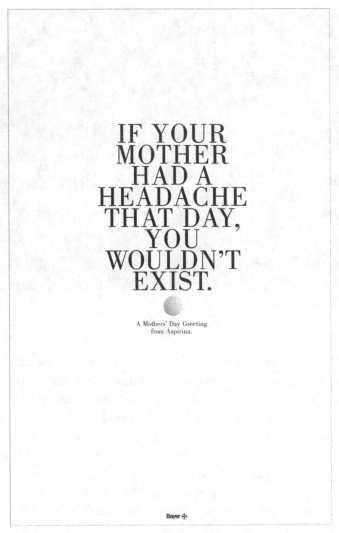

Figure 7.9   Aspirin.

and we know it. However, the question is whether we are going to consider taking the pill when we have a headache. This advertisement gives us an additional reason to do just that – albeit a humorous one. Emphasizing minor traits renew products in our eyes. When we thought we knew all there was to know about a product and the benefit it could bring to our lives – along comes the minor trait and we see the product in a completely new way.

When to use minor trait Inversion

Inversion of a minor benefit is not suitable to every product. Firstly, only use this tool if there is complete certainty that the audience is fully aware and appreciative of the product's primary benefits. If the product is not well enough known, using the Inversion tool on a minor benefit will miss its mark. Therefore, this tactic is appropriate generally for more mature products that are identified with their category. For example, the Inversion tool would work well on an advertisement for milk. The minor benefit of a glass of milk that was worked through the Inversion tool is expressed in Figure 7.10 in the tasty combination of milk with chocolate syrup.

Figure 7.10   Got milk? – Enhance with chocolate.

(a)   (b)

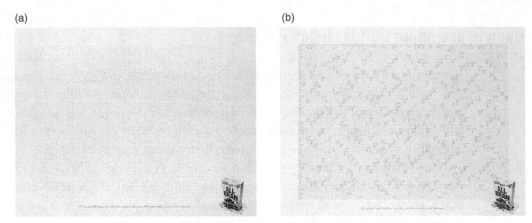

Figure 7.11   (a) A puzzle. Caption reads "If you don't have All-Bran every day, keep this to solve in the bathroom."
(b) A crossword. Caption reads "Not having All-Bran every day? Then look for the other 499 words while you're in the bathroom."

Generosity

An important tactic of Inversion is to show unlimited generosity, understanding and empathy (which of course does not rule out the use of compassion) towards those who do not use the product and do not enjoy its benefits. The advertisement does not judge and as stressed previously – it does not present a realistic view of what will happen to the viewer as a result of not having the product. It generously suggests assistance to those who don't enjoy the product's benefits. The tool does not express other methods to get the same result, but deals with the hardship that is the lot of the non-customer.

The advertisement in Figure 7.11 is generous and empathetic, even full of pity. Using the Inversion tool, the advertisement doesn't have even a hint of possible threat or fear. The advertisement shows great understanding to a consumer who does not use All Bran cereal and doesn't enjoy the benefits of the high percentage of bran fiber in the product. Given the missing benefit, the advertisement suggests ways of coping with the results.

When to use Inversion

When is it useful to use Inversion? Keep in mind that when using this tool, the product is not featured. The advertisement concentrates on the benefit that the product creates. Because of this, Inversion will not be the preferred strategy when advertising a product with benefits that are not very differentiated from their competition. Similarly, the brand name is often not central to the advertisement and thus Inversion should not be used for products or brands where this could pose a problem. Inversion is especially appropriate for products that are significant leaders in a category and will gain from promoting awareness for the entire category (Figure 7.12).

INSTRUCTIONS FOR USE

(1) Create a list of the central benefits or the minor benefits of the product. Remember that Inversion of a minor benefit is appropriate only for products that are recognized category leaders, with primary benefits that are well known to the market.

The product is an agency for buying media.

The central benefits of using the service offered by the agency: matching media to the target audience in the following ways:
- matching media to the place;
- matching media to the time;
- matching media to the advertising environment;

(2) Choose one benefit and try to compose a situation in which its absence could be problematic or ineffective. Do this for each of the product benefits.
- The advertisement is targeted to audience x but is shown at place that will only allow audience y to see it.
- The advertisement is relevant to time x but appears in time y.
- The advertisement appears in an environment that distorts its content, or does not give it legitimacy.

(3) Select specific content for the event you choose and make sure it's completely exaggerated and absurd. The event you choose should be something that will never occur (Figure 7.13).

And some for the road

(1) Inversion is suitable when the product leads its category and is identified with it. It is also important that the central benefits are well known to the target audience.

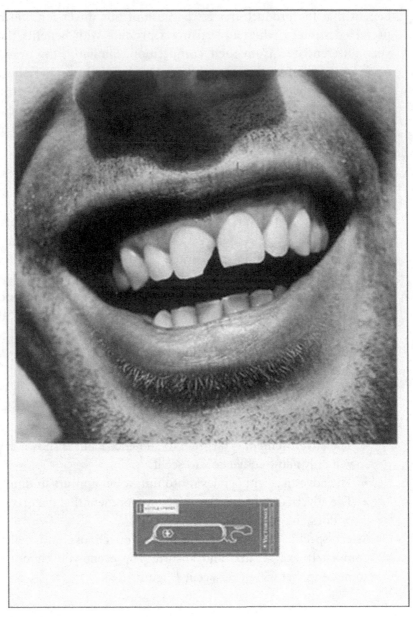

Figure 7.12   Tooth repair.

(2) In order to avoid vulgarity or bragging, one must exaggerate the shocking results of not owning the product to the extreme. In this case, extreme exaggeration actually helps you achieve greater believability regarding the message.

(a)

(b)

(c)

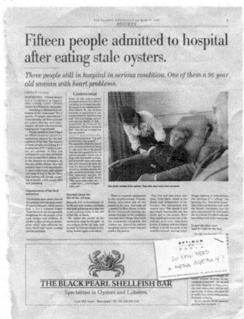

Figure 7.13 Ads in conflicting situations: (a) car; (b) chocolate; and (c) oysters

(3) It is good to remember the possibility of acting generously towards those unfortunates who don't have your product and offer them some sort of temporary aid with the suggestions in your advertisement. You don't provide them with the benefit that they would get by having the product – you just assist them in dealing with the fact that it's missing. You can also choose to use the Inversion tool on a minor benefit that is usually perceived as a trait on which there is no need to focus.

8 Extreme Effort

What to say when you have nothing to say

This last "extreme" tool is also the most common. Its popularity comes from a simple fact: exaggerating an effort is an efficient way to deliver a message about the product when there isn't actually anything significant to say about it.

If the product or the service in the advertisement is similar to its competition or possesses qualities that are well known to the market and there is no need to remind the audience about them again, the Extreme Effort tool can create the desired effect of keeping the product in the minds of the audience and making it stand out in a busy marketplace. The tool accomplishes this without emphasizing any particular quality. Instead, it emphasizes the overall attractiveness and desirability of the product itself.

How can a product's "overall attractiveness" be expressed? One way is to show the different steps consumers will be willing to take in order to protect their product (Figure 8.1).

Two types of scenes

There are actually two different applications of the Extreme Effort tool. Advertisements that employ this tool present an absurd scene. The scene will fall into one of two categories – either the company makes an extreme effort, or the client does.

(1) The company undertakes extreme and unreasonable efforts to create and /or to market the product in the best possible way, to draw the attention of the consumer, or to maintain his satisfaction or meet his needs. The actions are considered absurd because the investment made

Figure 8.1   Disaster averted.

in the product isn't commensurate with the price paid for it or the value of the product. Or similarly, the effort and investment in maintaining consumer satisfaction exceeds what is generally considered reasonable. This type of extreme can be suitable for service companies, which can demonstrate in an absurd way the lengths to which they are able to go in order to take care of their customers' needs (Figure 8.2).

A commercial advertising John West® canned fish begins with the pastoral image of bears hunting salmon at the edge of the river

(a)   (b)

Figure 8.2   (a) Pothole avoided and (b) tricycle transport.

Figure 8.3   Bear fight.

(Figure 8.3). When one bear is finally successful in his endeavor and throws the fish he just caught, a John West employee charges from out of nowhere and grabs it. The bear notices him and they begin to wrestle like two experienced fighters; the bear, however, is stronger and when it looks like it's a lost battle for the John West employee, he lands a well-placed kick to the bear's privates. While the bear is crumpled over in pain, the employee takes the fish and hightails it out of there.

(2) The customer or others in his environment undertake irrational actions in order to protect the product already in their possession or to get hold of the product. The characters presented in the advertisement invest a lot of money in its creation or in getting it; they are willing to sacrifice

valuable possessions or risk their well being in order to guard it or they choose it above something that is of hugely greater value. This type of use of the Extreme Effort tool appropriate for use with every type of consumer item. Its use is demonstrated in the examples of advertisements for Stella Artois beer (Figure 8.4): to get a taste of it, one is apparently prepared to sacrifice any number of other very valuable things.

When to use the tool

One of the "Got Milk" television commercials is situated in a plane. When the pilot gets chocolate chip cookies from the steward, he bites into one, filling his mouth – and immediately is desperate for some milk. He looks back to the body of the plane and notices a pitcher of milk resting on the service trolley, in order to be served to passengers. Unfortunately, the stewardess is nowhere to be seen and therefore, in order to get his desired milk, the pilot veers the plane in a downward angle so that the trolley will roll in his direction. Unbelievably, the trolley is stuck on a nut that had been carelessly thrown onto the passageway and impedes the trolley's progress. The pilot has no choice – he turns the plane to an impossible downward angle to get the trolley to pass over this obstruction. While the plane is in a dangerous downwards spiral, the trolley calmly makes its way towards the pilot. At the very last moment, when it seems that the goal has been accomplished, a door is opened to the bathroom right near the pilot and the trolley hits the man coming out of the bathroom with full force.

Extreme Effort is useful when planning an advertising campaign for a product about which there isn't something special to say. This is a category that is steadily growing. The more generic a product and the more difficult it is to differentiate between it and its competitors, the more the Extreme Effort tool becomes an accessible creative solution.

This tool does not provide a platform for delivering information and does not allow space for particular discussion of a product's traits or benefits, so it will most probably prove unsuitable to the advertising requirements of a new product or of an unfamiliar product that has a need for explanation. On the other hand, this type of advertising works especially well for products that are bought impulsively. To clarify, this tool is effective when the purchase decision for the product in the advertisement will not entail deep thought about the traits of the product or its advantages. Similarly, the Extreme Effort tool is recommended for use with products that are number

(a)

(b)

(c)

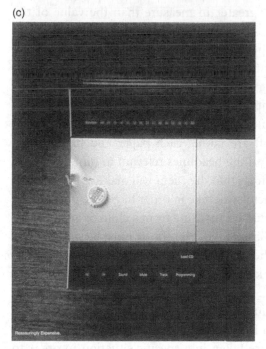

Figure 8.4 Stella Artois ads: (a) car; (b) guitar; (c) stereo.

two in a category and are not very differentiated from category leaders. The prime example of Avis® – We try harder – is of use of the Extreme Effort tool. Of course, the example shows the company making the extreme effort for the clients.

INSTRUCTIONS FOR USE

(1) Think of an unusual effort that the company can make in order to satisfy the client.

The product is a financial paper.

The message is that the information in the paper is very attractive to the client.

Possible unusual efforts are:

- The company can assign a well-heeled financial advisor to each client. The advisor will walk behind the client, following him everywhere – to business meetings, to the bathroom, to bed with his wife/husband or to vacation.
- The company plants undercover agents at the client's competitors' offices so that they will report on what is happening there.

(2) Think of an unreasonable company investment in the product or the service, which is significantly greater in measure than the value of the product or service.

- The faces of the paper's reporters are blackened with soot; they press their ears to the opening of rooftop furnaces, on buildings in which important business meetings are taking place.
- Divers dispatched by the paper raise a periscope to catch every word of what is happening on an important businessman's yacht.
- The paper's marketing staff are taking each paper and highlighting headlines per client, based on the headlines relevant to him.

(3) Think of special but absurd efforts that the client will make in order to get the product or that another factor will take in order to get the product for himself. The "other factor" could be anything from a child or an animal – all the way to a bank robber. Be creative. The efforts must be significantly greater and not in line with the value of the product.

- A group of robbers undertake an armed bank robbery, only to steal the paper.
- A tailored executive buys an unreasonable amount of fish in order to read the paper that the fish is wrapped in consecutively.

(4) Think of special efforts that are taken in a specific situation in order to protect or save the product – for example, situations in which someone

is threatening the client's hold on the product. The client fights gallantly for his product.

- A man sits in a park, on the banks of a river. He builds sophisticated "wind protectors" around himself in order to prevent the paper flying out of his hands.
- A man in a train crawls underneath his seat to read, so that no one will ask him for a piece of his paper.

Here are some situations in which the client has to sacrifice the product in order to achieve a certain goal, but the client chooses a different way to reach his goal or he abandons it.

- A man who is saved from a plane crash prefers to freeze from the cold, rather than use the paper to light a fire.
- A man whose children used the paper to line a cage of an exotic bird that he bought them for a birthday gift frees the bird in order to get the paper back.

(a)

(b)

(c)

(d)

Figure 8.5   Protecting the paper from (a) fish and chips; (b) fly swatting; (c) paint spots and (d) paint splashes.

- Armed men break into a house and ask the proprietor to open the safe. He refuses to do it because the paper is inside.

Here are some situations in which the client sacrifices something (a valuable object, the happiness of those around him) in order to keep the product, saving him from the need to give it away, use it or dispose of it.

- A man prefers to eat oily chips from his hat, rather than allow the paper to get dirty (Figure 8.5a).
- A woman hits an annoying fly with a valuable keyboard rather than use the paper that is close at hand (Figure 8.5b).
- A man lines the floor in the house he is painting with clothes and expensive carpets so that the floor won't get dirty – and neither will his precious paper (Figure 8.5c).
- A man paints his car and covers the car windows with personal diaries, pornographic magazines, or important documents (plane tickets, marriage licenses) to preclude using his beloved paper (Figure 8.5d).

And some for the road

(1) If there is nothing to say, it is best to say it with the Extreme Effort tool.

(2) The Extreme Effort tool allows for unlimited imagination and the implementation of absurd humor at its best, when you plan a scene within which the client is ready to do anything in order to get or to protect his product.

(3) In the event of service products, the exaggeration can emphasize outlandish efforts of the customer service department, of marketing people or of company managers in order to achieve client satisfaction.

9 Attribute–Value Mapping

The three questions

How do advertisers actually decide on the content of any given advertising campaign? When approaching any campaign, advertisers are driven by three basic questions:

(1) What to say?
(2) How to say it?
(3) How to clearly connect what was said with the product?

In providing different patterns of creative advertising, this book has focused on answering only one of those three questions – how to say it. The patterns provide advertisers with a type of creative direction; they offer advertisers a variety of options of how to get the message across in the most creative way.

The book also touches on the third question: how to clearly connect what was said with the product. In several instances we have explained how to connect the message with the product through the principle of Fusion. We also discussed methods of testing the message and making sure that it is connected with the product through the tools of Absurd Alternative and Extreme Consequence.

In this chapter, however, we will deal with the important and deceptively simple issue of: *What to say?* To help us navigate through what can often be quite a complex task we will follow a systematic process we call Attribute–Value mapping.

The process is kicked off by asking two very banal questions about the product.

(1) What is in the product?
    That is to say – what are the attributes of the product?
(2) What do I, the customer, get out of it?
    That is to say – what are the product's values?

Although these are two rather simple questions, don't let their simplicity fool you. Many organizations don't even bother to answer these questions in an organized way as part of their communications strategy, perhaps due to their apparent simplicity and obviousness. However, answering these two questions will provide all the information needed as background for the strategic planning of any advertising campaign or of any piece of marketing communication.

Getting the answers to these questions is central to deciding what to say. It helps us distinguish between a product attribute and a product value. Distinguishing between product attribute and value is the first step in devising a message that speaks directly to the target audience. For this to be completely clear, we need to explain a little about the difference between attribute and value.

The difference between an attribute and a value

Surprisingly, the distinction between a product attribute and a product value is not always that clear cut. Sometimes, even organizations that have been marketing their products for a long time don't have an unambiguous distinction with regards to product attributes and values.

The rule of thumb that determines if something is an attribute or a value is very simple: if the subject is the product, we are talking about a product attribute. If the subject is the customer, we are talking about a product value. Let us look into that a little more deeply.

An attribute is a characteristic of the product. It will be described in objective language and is always explained from the point of view of the organization. The product attribute does not provide an interpretation of the meaning of the product for the customer.

Some examples of attributes are listed here.

The jazz festival can be reached by traveling on a specific bus.

Antika mineral water comes from an ancient spring.

The policy of GT bike manufacturers is speed.

Canal Plus channel broadcasts the Wimbledon tennis tournament.

An attribute is usually measurable – but not always. The claim that Kodak®'s paper is of higher quality would be considered a product attribute, if only because the message is composed from the point of view of the company and not from the considerations of the customer. That is the central rule.

Value is the perceived benefit or gain that the customer expects to secure from the product or service, or from particular attributes of the product or

Table 9.1 Attributes and Values

Attribute	Value
It is possible to get to the Jazz Festival through a specific bus line.	I can easily get to the festival.
Antika mineral water comes from ancient springs.	I drink classic and selected water.
The policy of GT® is speed.	I buy fast bicycles.
Canal Plus® broadcasts the Wimbledon tennis tournament.	I can watch the tournament in real time.
The quality of the paper is high.	In order to get good pictures, I need to use Kodak® paper.

service. Notice that when we use the concept "value" it always means that the message is composed from the point of view of the benefit to the customer. Value never talks about anything from the point of view of the organization's vision. The value appears in first person (if I buy product X, Y will happen).

Some examples of value could be:
I will be a better parent if I drive a Volvo;
I will be taking care of my family if I buy life insurance;
(or, morbidly) I will die if I smoke cigarettes.

Thus, it is possible to fashion the following values from the attributes that were presented above (Table 9.1).

Added value of defining the value – the next step of the process: Beginning to seek new messages

Note the effect of composing the value in first person. It is not simply semantics. Certainly, first person speech has a certain impact on a consumer, but, in addition to that, composing the message in first person has great practical value to the advertiser as it allows him to reach deeper product insights. For example, if the attribute is "Air France flights allow rest" – it won't be very easy for the advertiser to devise new messages that relate to that. The idea of "comfort" can bring us to a dead end. When the starting point is "comfort," it will be difficult to produce a complete advertising campaign with new messages – that are different from each other.

However, when the message is composed in the first person, "I can spread out and rest in Air France flights," then we can more easily arrive at directions for new messages that express the personal meaning of this value: "I will arrive fresh and focused for meetings (that is to say, I will be a better manager); I deserve moments of rest (meaning, I maintain a balanced and rational life style); or I will invest in myself (which also means – I deserve the best)." Three completely different messages, one distinct value. Each of these messages could be directed at a different target group and for this reason, the first step in the process of tailoring messages to specific target groups is to distinguish between the qualities of the product and the values that are connected with them.

Later in the chapter, we will show how this example of pulling out different messages from a single value is the first step in creating a hierarchy of values that will help in deciding on what to say in the advertisement. However, let's go back to focusing on the list of a product's attributes and values.

It is generally sufficient to create an organized list of a product's attributes and values in order to discover the effectiveness of the distinction between the two. From our experience, marketing teams who work on the same product often have difficulty in agreeing on the values of a product and even on its attributes. Can we say that our service is fast? Is our product of better quality than its competitors? These are examples of issues that marketing teams discuss in order to compile their lists. The mere fact of raising these questions gives the advertiser a deeper awareness of appropriate messages and inappropriate messages. Thus, this process of defining attributes and values allows for renewed thinking on the entire advertising marketing strategy.

A few examples: Are these advertisements based on attribute or value?

In order to make the difference between an attribute and a value abundantly clear, see which stands in the center of the message that is delivered in the following three advertisements.

The first ad has an attribute at its center (Figure 9.1), driving the message: the air conditioner's quietness. The advertisement teaches us something about the product itself, and not about the meaning of this quietness for the customer. The second advertisement – for Volvo® – is an advertisement that discusses a product attribute through Metaphor, although without the use of Fusion (Figure 9.2). The car's physical toughness is represented, but this attribute does not directly and straightforwardly express the value of that attribute for the customer. Although the value clearly flows from the attribute, it is not presented in the advertisement itself. The third advertisement does

Figure 9.1   (a) Silence.

Figure 9.2   (b) Strength.

not teach us anything about the deodorant (Figure 9.3). It describes its value – sex appeal, which threatens the innocence of the nun in the picture.

Why don't people do what they know is right?

Ask any group of people dealing with marketing what should be delivered in marketing communications – value or attribute – and it is reasonable to

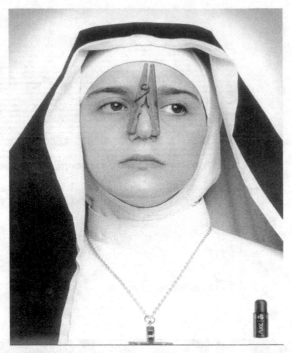

Figure 9.3   (c) Sensitivity.

assume that all will respond – value. It is common marketing wisdom that value should always be the focus of the message; the message of an advertisement should always focus on what interests the customer. This also allows for targeting advertisements to particular markets. However, if we analyze the advertisements that these same teams approved or came up with, many focus on a product attribute and not on the value to the customer. Throughout countless seminars with advertising people, we meet this paradox again and again: why do people who know that focusing on the value is the most important of all – in reality, create marketing communications that focus only on a product attribute?

There are a few responses to this question. Some of the reasons this occurs are "bad" and some are "good." On the bad side: it is easier and safer to talk about a product's attributes. Attributes are concrete, known, and defined. It is scarier to talk about value because the dangers in composing it are greater. Perhaps this value will not be perceived as valuable to every market segment. Maybe the value that the team composed is not significant enough for the potential customer. It is possible that the competitors will offer the same exact value. We may not have a strong enough way to back up the value we suggested with qualities of the product. All these reasons for not discussing product values in advertisements are fear based and therefore, we call them "bad" reasons. The fact that sometimes marketing teams find it hard to get out of their own head, is another "bad" reason. They are so wrapped up in the product and how great it is – they are so excited about what they created – that they forget to think about who is out there and how this product – and message – is going to affect them.

However, there are sometimes good reasons why marketing teams choose attributes as the focus of the marketing message. In some marketing cases, it is better to talk about an attribute. For example, if the market has already been educated to "get" the value immediately upon hearing about the product or product attribute, there is no need to explain the meaning of the product attribute. Here are some examples of attributes: "Vivisical hair products help in the growth of hair" or "Ikea's products are particularly cheap." Immediately upon hearing the product attribute, the value or benefit to the consumer is clear.

The connection between attribute and value

We have seen how important it can be for advertisers to give thought to the different attributes of the product and to distinguish these from the product values. However, it can be equally significant to test the symbiotic relationship between the two. As previously mentioned, the advertising message will usually focus on the value for the customer. Yet, the value is generally liberally interwoven with the attribute of the product. A significant task of marketing communication is to present this connection and to make sure that the connection is completely clear in the mind of the consumer. It is the role of the advertiser to make sure that the consumer knows that the benefit they desire will be found in the product's attribute.

However, there are some specific situations in which it is possible to present only one of the elements of the equation – either the product attribute or its value. We have outlined a number of these situations in the following list.

- If the attribute is already completely identified with the value to the customer, there is no need to present the value in a direct way. In this case, an explanation and emphasis of the attribute itself will be sufficient.

The consumer independently connects the attribute to the value to himself. When providing information on the Volkswagen® Polo's low fuel consumption, for example, advertisers need not worry about creating a clear message of the value of that attribute for the customer; the connection is already clear in the mind of the consumer. The value of low fuel consumption means "I will save money."

- At the same time, if the customer unequivocally views the value of the product as clearly dependent on its recognized attributes, it is not required to present the attribute. The Volvo® company has already educated its customers to identify the car with safety. Therefore, the company does not need to connect the value with a specific attribute of the car. However, please be clear that eliminating the attribute is only possible when there is complete certainty that the connection between the attribute and the value exists clearly in the mind of the consumer, as shown in Figure 9.4.

There are not that many situations in which the consumer automatically perceives a stable connection between the value and the attribute. When such a connection exists, as in the example in Figure 9.4, it is often the result of prolonged market education. Usually, marketing communication is focused on creating a stable connection between the attribute and the value. This is the focus because a connection of this sort that works creates a promise to the consumer that is at once differentiated from the market and is convincing.

Figure 9.4   Cotton wool protection from a car.

Notice, also, that the organization may perceive a different connection between a product attribute and its value than the connection perceived by the audience. The computerized brake system may be the central element in the safety of a car, but few customers know this. Even if the existence of micro-processor X867 in a video machine makes it very easy to operate, most customers are not aware of this.

Sometimes the organization is not aware that there is not an immediate connection in the mind of the consumer between the attribute and the value. For example, the attribute that a bank's branches are spread out throughout the country is not necessarily connected in the mind of the customer as the value – there is a branch near my house. In these cases, one of the goals of the advertisement is to create that connection in the mind of the consumer.

In some situations, the marketing team has to call on their creativity simply to find a value for a product attribute. Cases that require deep thought include product attributes that are taken for granted by the audience, or attributes that actually negate values. For example, Orbit® gum reduces the level of acidity in the mouth – and therefore it protects the health of the teeth. This message is so identified with Orbit® gum that it would come as a colossal surprise to the consumer if they were aware that simply chewing paper would reduce the level of acidity in the mouth. In effect, any time that we put something in our mouths saliva is created and it is this biological process that reduces the acidity in the mouth. The Orbit® marketing team, therefore, discovered a general pro duct trait that was taken for granted (the product is meant for chewing) and they created a value around this attribute that became identified conceptually with the product. The Orbit® customer is identified with this value – Orbit® chewers do not consume gum because of the taste, they chew gum to maintain the health of the mouth and the gums, especially after meals.

Distinguishing between attribute and value levels

We will immediately clarify ways to make a novel connection between the value and the attribute that will assist in designing the advertising message. However, it will help to first distinguish between different levels of attributes and values.

- Attributes are usually made up of more basic attributes. A car with a high safety standard was awarded this standard because of a tough bumper, improved air bags, a strong body and so on. How do we discover these other attributes when looking at a product? First, start with basic attributes of the product – and then ask "Why?" Why is the car safe? Because it has an ABS brake system, because it has a strengthened body and so on. Doing this for each of the attributes helps you to begin to map out the various product attributes.

- Values can also be perceived in this hierarchical fashion. The value of "safety" can be explained by wider perceptual values, connected to the customer's thoughts about "I take care of my family," "My feeling of personal security in life is important to me." These second realm of values arrived at through the value of "security" are considered "higher values." Notice that this higher level is less trivial, less general in its scope. This allows us to differentiate the product's promise from the competition. In order to arrive at this "higher" level of value, we ask the question, "So what?" An example of this would be: so what if the car is safe? So I will live more, I will be a better parent, I take care of the people that I love or I replace my car only when I am sick of it.

The chain of hierarchy can be presented in a diagram that we call the Attribute–Value map (AV map) (Figure 9.5).

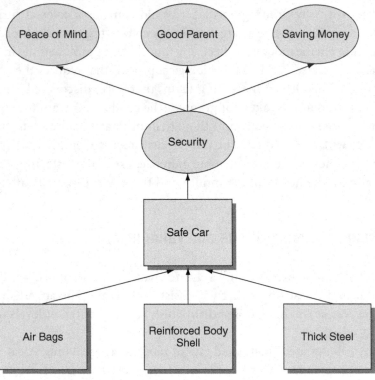

Figure 9.5   Attribute–Value mapping.

What the AV Mapping offers

The AV map helps us consider various promises and values of the product. It actually helps us decide what to say in our advertisement. Through the map we can test how strong these promises really are, and if they are based in actual facts and in the reality of the product or service. This tool allows us to distinguish the following elements.

(1) To understand what the customer gains from the product and to compose what we would like him to perceive as a benefit.

(2) To connect the product and its attributes directly and clearly to the values that have meaning for the customer.

(3) To distinguish different target groups and to tailor a relevant profile of the product and its values for each one of them – as is relevant for them.

These objectives can be achieved by critically assessing the system of attributes and values you uncover and by identifying weak spots that exist in the system.

The following are examples of weak spots that must be addressed if discovered in a mapping exercise.

(a) Values that are not backed by attributes. We tend to promise something but can't exactly say on what we are basing the promise. We can say that yogurt promises excitement and fun in the same way that we can say that yogurt promises health. However, each of these promises will demand a different type of supportive attribute. The first message would require attributes that strengthen the concept of excitement and fun. This could be through the packaging, color, touch, etc. The second value would be supported by attributes that touch on the nutritional value of the product. Mapping of this kind can point to the need in future development of the product or service. If there is a value that doesn't have strong attributes connected to it – the organization can undertake to develop new attributes to support the value.

(b) An attribute that does not provide a value. Once we discover a product attribute that we haven't been able to convert to a value, we are open to new possibilities in our marketing communication. We can create a new value for this attribute! Examples of this would be air conditioning that also purifies the air or a chain of restaurants that hires young men and women from broken homes as waiters and assistant chefs.

(c) Negative connections between the attributes and the values of the product. Ice cream is fattening, a fact that can raise negative values for specific customers. In these instances, mapping provides a check of these negative values and allows the advertiser and product manufacturer several avenues to handle the difficult situation: they can choose to improve the product, as in the introduction of dietetic ice cream for a specified target group; or they can choose to turn the problem into a solution. For example, offer particularly rich ice cream, a product that allows the customer to enjoy all that life has to offer and to go all the way.

How to build your own Attribute–Value map

Map out all the product attributes and values

In the first stage, make a list of each of the attributes of the product and the existing values that are connected to them, as perceived by the company today.

We will use the example of an organization that markets natural juices. Find the main attributes mentioned by the organization's marketing team: The juice is a naturally-squeezed product; the product is rich in vitamin C; the product comes in many variations with different tastes; the product's packaging is orange in color; the product is made from natural ingredients only.

The values that are central to this advertising strategy are: I get a healthy product; the taste surprises me every time anew; I think the product is tastier than its competitors (Figure 9.6).

We will organize this data spatially, as described in the next section.

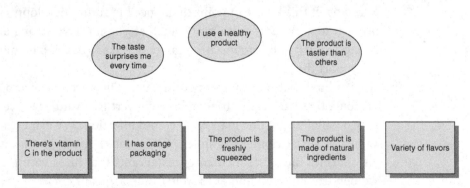

Figure 9.6   Listing Attributes and Values.

Map at high resolution

After initially mapping out the attributes and values of the product, use these existing attributes and values to arrive at a higher resolution of mapping: mapping the additional attributes and values of the product, even those that seem to be secondary, as well as more basic product attributes that support the attributes already mentioned.

(a) We begin at the middle with the central attributes of the product or the service.

(b) We move up to "higher" attributes by asking questions like "What will I get out of it"? It is important to go as high as possible in the search for attributes and values, because it is at these higher levels that non-trivial messages can be found. Thus, for example, when the value "health" is raised with regard to a product, we continue with "So what? What will I get out of it that the product is healthy?" The answers can come from several directions and will open the door to discovering higher values such as, "I give my children healthy products." Which leads to " I am a better parent" or on the flip side, "I don't compromise on what I deserve."

(c) The mapping must also continue in the other direction – downwards. We get to this through the question "Why?" Based on what can I say to show that the attribute which I mentioned exists in the product? For example, if we ask about the attribute "a variety of flavors," asking "why" we discover an even more basic answer "every quarter a new flavor is released."

Figure 9.7 demonstrates how to compose additional values that are derived from existing values and how to define the attributes of the product.

Making connections

After getting all the attributes and values down on paper and completing the widest possible picture of the product, we boldly draw arrows that connect between the values and the attributes from which they spring up. Draw additional arrows between the basic values and the higher values that are derived from them (Figure 9.8). Notice that simply creating the arrows creates a clustering of attributes and values in close groups. Each of these groups suggests the possibility of a different advertising message.

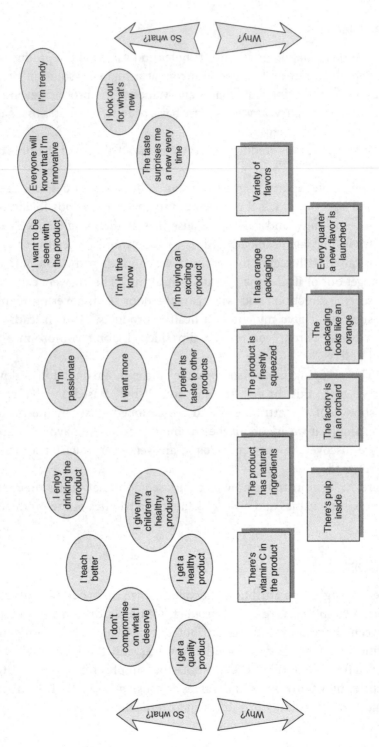

Figure 9.7 Composing additional values and their attributes.

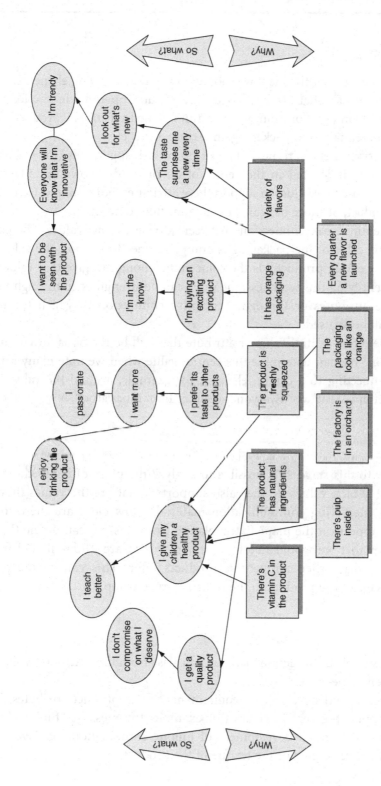

Figure 9.8 Linking Attributes and Values.

Analysis of the map

When analyzing the map, we should review the following elements.

- Any values that are not based on attributes open a window for product development. For example: "it is fun for me to drink the product" forces a differentiation in packaging, in texture, taste, etc.
- Attributes that are not followed up by existing values (for example, the factory is located in the center of an orchard) allows the advertiser to compose new values to support these attributes. For example: I get a natural product, or more basically, I am connected to nature.
- Identify weak connections between attributes and values. For example, "the color of the packaging is orange" comes to support the value "I buy an exciting product." Is the color alone enough to give that value? Upon identifying this weakness, the value can be strengthened through supportive attributes or through a cognitive connection in which the message will be focused.
- Identify a central value or attribute that will be the basis for communication. We can determine this central value when we note many arrows all connecting to a certain element; for example, around the value "I am a better parent" or "Everyone sees that I am innovative."

Choice

Prior to this stage, it is possible to analyze the place of competitors on the map: Which values do they also support? What are the values they use in their marketing communication material? How close are these to values that appear on the map? Extrapolating from this, we can define the higher values that we would like to focus on, which are differentiated from the values that other competitive products offer. This can be the potential message of the product to the specific target group.

Action items

An example of the action plan that can arise from the Attribute–Value map is presented here.

- Define and develop the requirements of the product attributes that will support the central values (for example, the message "fun" will be supported through development of unique combinations of flavors, design of the packaging or the texture of the product.

- Choose a new promise and test this promise in the consumer market.
- Think about implementing the message that you chose in an advertising campaign, by using one of the creative tools discussed in the book.

And some for the road

(1) Mapping attributes and values allows for renewed thinking about the message chosen for the product or the service. It forces us to check if there are attributes that are not connected to an existing message and to connect the product value to higher values that a customer gets. However, at the same time, looking at values that are not connected to attributes provides the opportunity to develop new attributes for the product.

(2) Mapping attributes and values allows for strategic niche marketing. It points to the condition of the product today as opposed to its competitors and points to possible directions for product development. Don't fear messy and busy maps. A map filled with information will increase the chances of locating a greater pool of new and successful potential ideas

(3) Asking the questions "So what? and "Why?" seems simple – maybe even too simple. But thinking through these questions about the values and existing attributes can lead us to unexpected, newer, deeper, and more meaningful breakthroughs and marketing messages.

Photograph Credits

I.1   Used with permission from Euro RSCG Singapore.
1.1   Used with permission from Abbot Mead Vickers BBDO, London.
1.2   Used with permission from Cramer–Krasselt, Chicago.
1.3   Used with permission from Z. Publicidade, Lda.
1.4   Used with permission from The Chupa Chups Group.
1.5   Used with permission from Amsterdam Advertising. Art director: Darre van Dijk, Copywriter: Piebe Jan Piebenga, Photographer: Fulco Smit Roeters.
1.6   Used with permission from Forsman & Bodenfors, Gothenburg.
1.7   Used with permission from careerbuilder.com.
1.8   Used with permission from Ullern Basketball Club.
1.9   Used with permission from Liceo Cultural Britanico.
1.10 Used with permission from McCann–Erickson, the Netherlands.
1.11 Used with permission from Acordia (Wells Fargo Insurance Services).
1.12 Used with permission from Lowe Lintas, UK.
1.13 Used with permission from Land Rover SA.
1.14 Used with permission from Abbot Mead Vickers BBDO, London.
1.15 Used with permission from The Auckland Regional Council.
1.16 © Greenpeace.
1.17 Used with permission from SHELTER, UK.
1.18 Used with permission from The Palm Springs Bureau of Tourism.
2.1   Used with permission from FCB, Portugal.
2.2   Used with permission from Contrapunto, Spain.
2.3   Used with permission from The Cancer Council, Australia.
2.4   Used with permission from Sonnenburg Murphy Leo Burnett, Johannesburg.
2.5   Used with permission from Grey Worldwide, Belgium.
2.6   Used with permission from Leo Burnett, Argentina.
2.7   Used with permission from Allanson Nilsson Riffi BBDO, Gothenburg.
2.9   Courtesy of DaimlerChrysler Corporation.
2.10 Used with permission from Saatchi & Saatchi, Sydney.
2.11 Permission for use granted by Network Equipment Technologies, Inc.

2.12   Used with permission from Lowe Bull Calvert Pace, Cape Town.
2.13   Used with permission from Saatchi & Saatchi, Cape Town.
2.14   Used with permission from Amnesty International, Spain.
3.1   Used with permission from ASH Scotland.
3.2   Used with permission from Minnesota Zoo.
3.3   Used with permission from McCann–Erickson Worldwide.
3.4   Used with permission from Head NV.
3.5   Bally Shoe (Wieden and Kennedy, 1992, The One Show).
3.6   Used with permission from BBDO Bangkok.
3.7   Used with permission from Pirella Gottsche Lowe, Milan.
3.8   Used with permission from Tiempo/BBDO, Spain.
3.9   Used with permission from BMP DDB, London.
3.10   Used with permission from Campina.
3.11   Used with permission from Vinizius/Young & Rubicam, Spain.
3.12   Used with permission from The South African Bottled Water Association.
3.13   Used with permission from Vinizius/Young & Rubicam, Barcelona.
3.14   © The LEGO Group.
3.15   Nike Air (Wieden and Kennedy, 1992, The One Show).
4.1   Used with permission from the WWF, Australia.
4.2   Used with permission from Euro RSCG Singapore.
4.3   Used with permission from Daffy's; Creative by DeVito/Verdi Advertising, NYC.
4.4   Reproduced by Special Permission of *Playboy* Magazine. Copyright © Playboy.
4.6   Used with permission from TBWA Chiat Day, New York.
4.7   Used with permission from IKEA; Creative by Honegger/von Matt.
4.8   Used with permission from General Mills, Inc.
4.9   Used with permission from McCann-Erickson Worldwide.
4.10   Used with permission from Euro RSCG Singapore.
5.1   Used with permission from Philips.
5.2   Used with permission from Ogilvy & Mather, New Delhi.
5.3   Used with permission from Tiempo/BBDO Spain.
5.4   Used with permission from Hill Holliday Connors Cosmopulos.
5.5   Used with permission from Crispin Porter + Bogusky, Miami.
5.6   Used with permission from Saatchi & Saatchi, Argentina.
5.7   Used with permission from DM9DDB Publicidade, Sao Paulo. Photographed by Manolo Moran.
5.8   Used with permission from Sara Lee Corporation.
5.10   Used with permission from Miller Brewing Company, Milwaukee, WI.
5.11   Used with permission from Almap/BBDO, São Paulo.

6.1	Used with permission from Craig Cutler Studio.
6.2	Used with permission from Abbot Mead Vickers BBDO, London.
6.3	Used with permission from Bauman Ber Ribnai, Tel Aviv.
6.5	Used with permission from Leo Burnett, Warsaw.
6.6	Used with permission from Euro RSCG Singapore.
6.7	Used with permission from McCann-Erickson, Budapest.
7.1	Used with permission from Publicis Casa de Vall Predreno & PRG, Madrid.
7.2	Used with permission from VISUAL – directed by ENJOY SCHER LAFARGE.
7.3	Used with permission from McCann Erickson.
7.4	Used with permission from TBWA, Singapore.
7.5	Used with permission from Contrapunto, Madrid.
7.6	Used with permission from the Sara Lee Corporation.
7.7	Used with permission from Network BBDO, Johannesburg.
7.8	Used with permission from Group T.
7.9	Used with permission from Almap/BBDO, São Paulo.
7.10	Used with permission from Goodby Silverstein & Partners, San Francisco. Photographer: Jack Anderson.
7.11	Used with permission from Leo Burnett, São Paulo.
7.12	Used with permission from Mullen Advertising, Wenham, USA.
7.13(a)(b)	Used with permission from Vitruvio/Leo Burnett, Warsaw.
7.13(c)	Used with permission from OMD Europe.
8.1	Used with permission from Carmichael Lynch, USA.
8.2	Used with permission from CLM/BBDO – Issy-les-Moulineaux.
8.3	Used with permission from Leo Burnett, London.
8.4	Used with permission from Lowe Worldwide.
8.5	Used with permission from Lowe Bull Calvert Pace, Cape Town.
9.1	Used with permission from BGH.
9.2	Used with permission from Pirella Gottsche Lowe, Milan.
9.4	Used with permission from Abbot Mead Vickers BBDO, London.

Index